Monographs on Astronomical Subjects: 8

General Editor, A. J. Meadows, D.Phil.,

Professor of Astronomy, University of Leicester

Early Emission Line Stars

In the same series

1 The Origin of the Planets
I. P. Williams

2 Magnetohydrodynamics
T. G. Cowling

3 Cosmology and Geophysics
P. S. Wesson

4 Applications of Early Astronomical Records
F. R. Stephenson and D. H. Clark

5 The Nature and Origin of Meteorites
D. W. Sears

6 Cosmic X-ray Astronomy
D. J. Adams

7 The Isotropic Universe
D. J. Raine

Early Emission Line Stars

C. R. Kitchin
Hatfield Polytechnic Observatory

Monographs on Astronomical Subjects: 8
Adam Hilger Ltd, Bristol

British Library Cataloguing in Publication Data

Kitchin, C. R.
Early emission line stars. – (Monographs on astronomical subjects, ISSN 0141-1128; 8)
1. Early stars
2. Spectrum, Solar
I. Title II. Series
523.8′7 QB843.E2

ISBN 0-85274-402-1

Published by Adam Hilger Ltd, Techno House, Redcliffe Way, Bristol BS1 6NX.
The Adam Hilger book-publishing imprint is owned by The Institute of Physics.

Typeset by Unicus Graphics Ltd, Horsham, and printed in Great Britain by Pitman Press Ltd, Bath.

I wish to dedicate this book to
my wife
Christine Elizabeth
for her patience and support
during the writing of it

Preface

Many types of stars contain emission lines in their spectra. Some only have one or two emission lines in an otherwise normal spectrum. Others have spectra which are comprised of nothing but emission lines. Many more stars which appear with pure absorption spectra in the optical spectrum are found to have emission lines when they are observed in the ultraviolet. The presence of bright lines in the spectrum of a star signifies a breakdown of the normal assumptions made about a star's atmosphere. They indicate an object which is peculiar in the sense that it differs in some way from a standard main sequence star. It is human nature to be interested in the out-of-the-ordinary, and even astronomers are human! A highly disproportionate amount of effort for the number of stars which are involved has therefore been devoted to the study of emission line stars of all types.

The aim of this book is to examine some of this work (the techniques used, the results obtained and the models inferred) which applies to the stars and stellar systems at the hot end of the temperature range. The defining criteria for the inclusion of a star in the book is therefore a high temperature (10 000 K or higher, spectral types A, B, O) and the presence of emission lines in the optical spectrum. The resulting selection of types of object may seem somewhat disparate at first sight, but closer examination will show a number of recurring themes, and surprisingly similar models for many of the stars. In particular, in almost all cases the emission arises from a region around the star where thin hot gas glows brightly against a dark background. Mass loss, at rates which are significant on an evolutionary timescale, is very frequent and usually arises as a radiatively driven stellar wind. In some cases the mass loss is so large as to reduce the star to zero mass in as little as 1500 years! Interacting binary systems are often invoked as models for the systems, and the emission lines are postulated to arise in the accretion stream between the stars. Most of the systems are post-main sequence objects with many of them about to become white dwarfs. In all cases the emission lines indicate that the star is at a crucial stage of its evolution, and they act as signposts to call the attention of astronomers to an object worthy of detailed study.

C. R. Kitchin
Hatfield Polytechnic Observatory

Contents

1. Extended Atmospheres and the Origin of Emission Lines

The presence of an emission line in the spectrum of a star is almost invariably the sign of a very extended atmosphere, or of a nearby gaseous region physically associated with the star. In such a situation the common assumptions made in the modelling of stellar line producing regions (plane-parallel atmosphere, principle of detailed balancing, local thermodynamic equilibrium and even radiative equilibrium) have to be modified or abandoned. Of the several mechanisms for the production of emission lines, those found in practice to be most important are variations on the Rosseland cycle. In these, for a variety of reasons, a single absorption of short-wavelength radiation followed by several emissions at longer wavelengths becomes more probable than the reverse process. Hence emission lines occur at the lower frequencies.

The Schuster mechanism, whereby scattering and absorption can produce emission or absorption lines depending upon the temperature gradient, is the only way in which emission lines can arise in the integrated spectrum of a star whose atmosphere approximately fits the usual assumptions. This process may be important in OB supergiants where Thomson scattering is significant. At the other extreme, emission lines can be produced by the breakdown of radiative equilibrium. For example, in the Solar Corona, acoustic energy leads to the production of emission lines, and in novae, acoustic energy or energy from short-lived radioactive nuclei may lead to the enhanced production of [Fe VII] emission. Finally, motions within the atmosphere may redistribute the flux to produce emission components to absorption lines, as in some P Cygni profiles.

1.1. The Rosseland Cycle

1.1.1. *The Basic Cycle*

The radiation at any point except deep inside the star is strongly anisotropic. A commonly used measure of the degree of anisotropy is the dilution factor, W, defined as the proportion of the sphere centred on the point of interest which is occupied by the star:

$$W = \omega/4\pi, \tag{1.1}$$

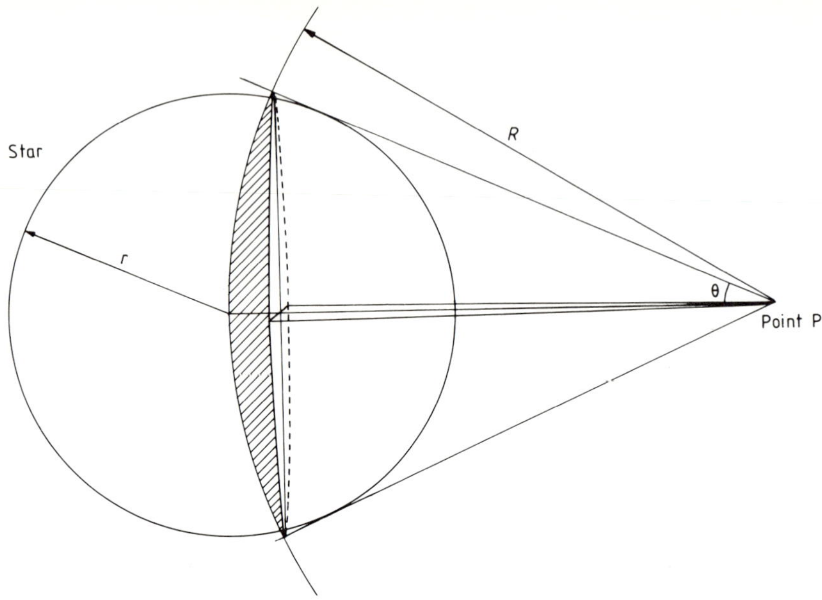

Figure 1.1. Dilution factor.

where ω is the solid angle subtended at the point of interest by the stellar disc. In figure 1.1 the fraction of the sphere centred on P which is intercepted by the star is shaded. For a stellar radius, r, and a distance R of P from the centre of the star, we have

$$\theta = \sin^{-1}\frac{r}{R}, \tag{1.2}$$

giving the area of the shaded region as

$$\text{Area} = 2\pi R\left[R - R\cos\left(\sin^{-1}\frac{r}{R}\right)\right] \tag{1.3}$$

$$= 2\pi R^2\left[1-\left(1-\frac{r^2}{R^2}\right)^{1/2}\right] \tag{1.4}$$

whence

$$\omega = 2\pi\left[1-\left(1-\frac{r^2}{R^2}\right)^{1/2}\right] \tag{1.5}$$

and

$$W = \frac{1}{2}\left[1-\left(1-\frac{r^2}{R^2}\right)^{1/2}\right]. \tag{1.6}$$

When $r^2/R^2 \ll 1$,

$$W \simeq \frac{1}{4}\frac{r^2}{R^2}. \tag{1.7}$$

The value of W ranges from 0.5 at the surface of a star to 10^{-15} in large planetary nebulae. From the definition of the dilution factor, it is clear that the radiation energy density at the point P, is simply reduced by a factor W from its value just below the surface of the star. For a black body

$$u(\nu) = WU(\nu, T_{\text{eff}}) \tag{1.8}$$

where $u(\nu)$ is the energy density at P, frequency, ν, and $U(\nu, T_{\text{eff}})$ is the black body energy density at frequency ν, and temperature T_{eff}.

The basic Rosseland cycle (Rosseland 1929) considers a three-level atom (see figure 1.2) and the two absorption-re-emission cycles; 1-3-2-1 (cycle A), and 1-2-3-1 (cycle B). Normally these are assumed to occur equally frequently (principle of detailed balancing). However, if the atoms are illuminated by dilute radiation, then two absorptions are required in cycle B, compared with only one in cycle A, and cycle A becomes more probable than B by a factor equal to the reciprocal of the dilution factor (as shown below).

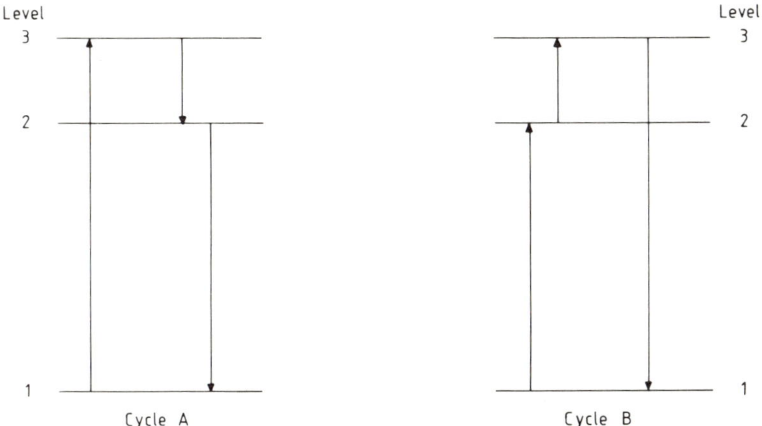

Figure 1.2. Rosseland cycle in a three-level atom.

The number of absorptions between two levels p and q, N_{pq}, is given by

$$N_{pq} = u(\nu_{pq})\, B_{pq} n_p \tag{1.9}$$

$$= n_p B_{pq} W U(\nu_{pq}, T_{\text{eff}}), \tag{1.10}$$

where B_{pq} is the Einstein absorption probability and n_p is the population of atoms in the pth level. The proportion of electrons in the third level of our three-level atom which undergo a transition to the second level is

$$P_{32}=\frac{A_{32}}{A_{32}+A_{31}}, \tag{1.11}$$

where A_{pq} is the Einstein spontaneous emission probability from level p to level q (stimulated emissions are ignored because the radiation is dilute). The proportion jumping from level 2 to level 1 is

$$P_{21}=\frac{A_{21}}{A_{21}+B_{23}WU(\nu_{23},\,T_{\text{eff}})}, \tag{1.12}$$

so that the number of cycles of type A is

$$N_{\text{A}}=n_1B_{13}WU(\nu_{13},T_{\text{eff}})\frac{A_{32}}{(A_{32}+A_{31})(A_{21}+B_{23}WU(\nu_{23},T_{\text{eff}}))} \tag{1.13}$$

$$=\frac{B_{13}A_{32}A_{21}}{(A_{32}+A_{31})(A_{21}+B_{23}WU(\nu_{23},\,T_{\text{eff}}))}n_1WU(\nu_{13},T_{\text{eff}}). \tag{1.14}$$

Similarly the number of cycles of type B is

$$N_{\text{B}}=\frac{B_{12}B_{23}A_{31}}{(A_{21}+B_{23}WU(\nu_{23},\,T_{\text{eff}}))(A_{31}+A_{32})}n_1W^2U(\nu_{12},T_{\text{eff}})\,U(\nu_{23},T_{\text{eff}}), \tag{1.15}$$

giving

$$\frac{N_{\text{A}}}{N_{\text{B}}}=\frac{1}{W}\frac{B_{13}U(\nu_{13},T_{\text{eff}})}{A_{31}}\frac{A_{32}}{B_{23}U(\nu_{23},T_{\text{eff}})}\frac{A_{21}}{B_{12}U(\nu_{12},T_{\text{eff}})}. \tag{1.16}$$

Now,

$$A_{pq}=\frac{g_q}{g_p}\frac{8\pi h\nu_{pq}^3}{c^3}B_{qp} \tag{1.17}$$

and

$$U(\nu_{pq},T_{\text{eff}})\simeq\frac{8\pi h\nu_{pq}^3}{c^3}\exp\left(\frac{-h\nu_{pq}}{kT_{\text{eff}}}\right) \tag{1.18}$$

(Wien distribution), so that

$$\frac{A_{pq}}{B_{qp}U(\nu_{pq},T_{\text{eff}})}=\frac{g_q}{g_p}\exp\left(\frac{h\nu_{pq}}{kT_{\text{eff}}}\right) \tag{1.19}$$

$$=\frac{n_q'}{n_p'} \tag{1.20}$$

(by Boltzmann's equation), where n'_p and n'_q are the local thermodynamic equilibrium (LTE) populations of levels p and q. Hence

$$\frac{N_A}{N_B} = \frac{1}{W}\frac{n'_3 n'_2 n'_1}{n'_1 n'_3 n'_2} \tag{1.21}$$

$$= \frac{1}{W} \tag{1.22}$$

So that cycles of type A are $1/W$ times as frequent as cycles of type B.

For points more than a few stellar radii away from the star, type B cycles effectively cease, leaving only type A. The radiation is then 'degraded' by being converted to lower frequencies. Very strong emission lines can result at the lower frequencies when the initial absorption takes place in an ultraviolet emission line which is optically thick ($W = 1$). (The measurement of the intensity of optical lines resulting from this process, leads to a very long base-line colour temperature known as the Zanstra temperature, which is discussed in detail in chapter 3.)

Extension to more complex atoms than the three-level atoms shows that the populations of normal higher states relative to the ground state are reduced by a factor W (Wellman 1955). Metastable states, however, are affected by dilute radiation in the same way as the ground state, and their relative populations are substantially unchanged from their undiluted values.

1.2. Fluorescence

The special case when the original absorption in a type A cycle is an ionisation is known as fluorescence, or ionisation and recombination. The emitted spectral lines have a characteristic intensity distribution. For hydrogen, the intensity ratios in the optical spectrum are known as the Balmer decrement, and may be calculated (see table 3.1). For a nebula which is optically thick in the Lyman lines and continuum, every Lyman photon (except those of Lyα) eventually results in Lyα, Balmer and possibly photons from higher series (see figure 3.23).

1.3. Selective Fluorescence

This process (sometimes also known as Bowen resonance fluorescence) does not depend directly on the presence of dilute radiation for its operation. Bowen (1934, 1935), pointed out that emission lines of doubly ionised oxygen from the 3d $^3P_2^0$ level were present in planetary nebulae, while other lines which might also be expected to be present were not visible. He showed that this arose from the overpopulation of the O III 3d $^3P_2^0$ level due to absorption in the He II emission line at 30.378 nm by the O III 30.382 and 30.362 lines (figure 1.3). About half of the He II Lyα photons are degraded by this mechanism into O III lines in a typical

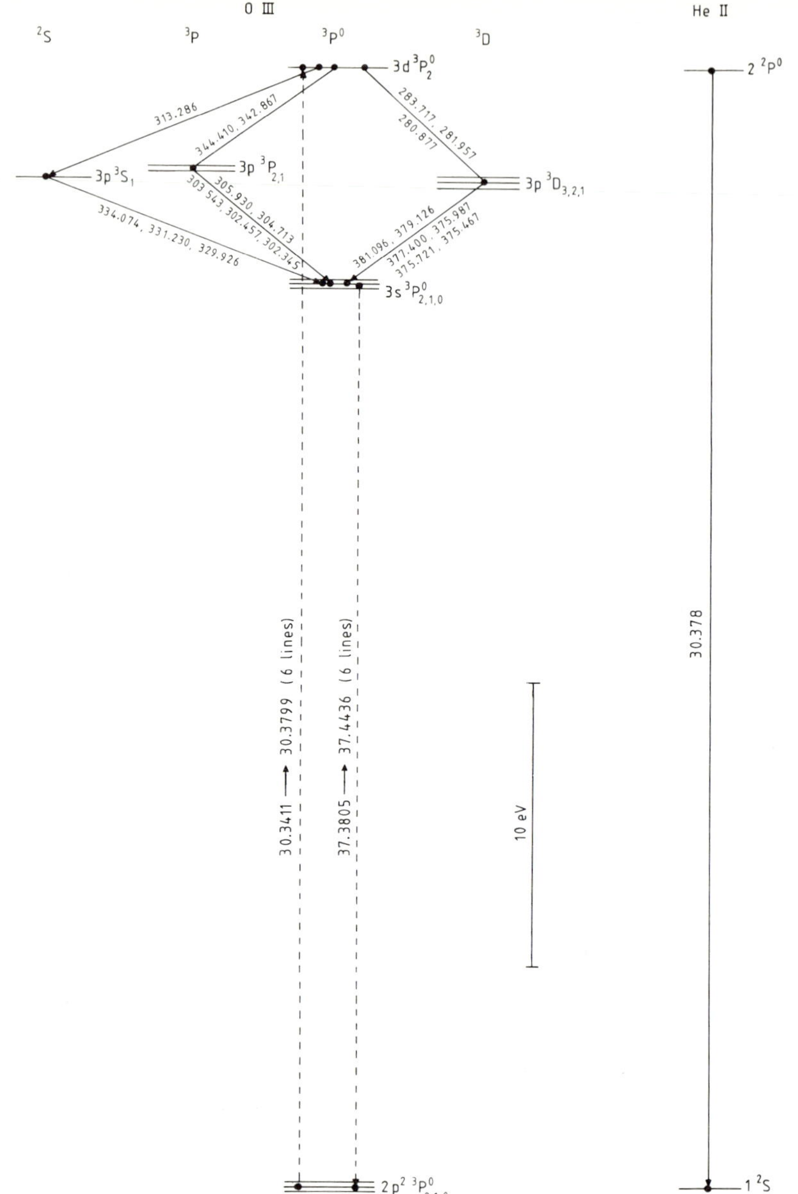

Figure 1.3. Partial Grotian diagram of O III and He II showing the Bowen resonance–fluorescence cycles.

nebula. A number of atoms have lines close to the ultraviolet resonance lines of common elements, and may therefore be candidates for this process. Another example is the O I 844.6 line in Be stars which arises from a coincidence of an O I line with Lyβ (Kitchin and Meadows 1970). Lines of N III and C III in Of stars, and lines of Si IV in P Cygni may also be attributable to selective fluorescence.

1.4. Schuster Mechanism

The ratio of line flux to continuum flux, R_ν, may be written (Mihalas 1978, Schuster 1905):

$$R_\nu = F_\nu/F_c \tag{1.23}$$

$$= \frac{[b/(1+\beta_\nu)] + [3\epsilon\beta_\nu/(1+\beta_\nu)]^{1/2}a}{1+[\epsilon\beta_\nu/(1+\beta_\nu)]^{1/2}} \frac{1+[1-\sigma_\nu/(\kappa_\nu-\sigma_\nu)]^{1/2}}{b+a\{3[1-\sigma_\nu/(\kappa_\nu-\sigma_\nu)]\}^{1/2}} \tag{1.24}$$

where β_ν is the ratio of line to continuum opacity, ϵ is the (classical) fraction of the line emission which is thermal in origin, σ_ν is the scattering coefficient, κ_ν is the absorption coefficient and a and b define the temperature gradient,

$$B_\nu = a + b\tau,$$

where B_ν is the Planck function and τ is the optical depth.

If

$$\frac{\sigma_\nu}{\kappa_\nu + \sigma_\nu} = 1$$

(i.e. the absorption is insignificant when compared with the scattering; such a situation may arise in early supergiants where Thomson scattering by free electrons may predominate), then,

$$R_\nu = \frac{1/(1+\beta_\nu) + (a/b)[3\epsilon\beta_\nu/(1+\beta_\nu)]^{1/2}}{1+[\epsilon\beta_\nu/(1+\beta_\nu)]^{1/2}}. \tag{1.25}$$

When $\epsilon = 0$ (pure scattering)

$$R_\nu = \frac{1}{1+\beta_\nu} \tag{1.26}$$

and the line is always in absorption. However, if $\epsilon = 1$ then

$$R_\nu = \frac{1/(1+\beta_\nu) + (a/b)[3\beta_\nu/(1+\beta_\nu)]^{1/2}}{1+[\beta_\nu/(1+\beta_\nu)]^{1/2}}. \tag{1.27}$$

Since β_ν is normally much greater than unity, we may write:

$$R_\nu \simeq \left(\frac{\sqrt{3}}{2}\right)\left(\frac{a}{b}\right) \tag{1.28}$$

and the line will now be in emission or absorption depending upon the temperature gradient. The shallower the temperature gradient ($b \ll a$), the brighter the emission. Detailed consideration of the process (Schuster 1905) shows that the line will be completely in emission, with a double emission line profile, for

$$a/b \geqslant 2/\sqrt{3}, \tag{1.29}$$

and completely in absorption for

$$a/b \leqslant 1/\sqrt{3}. \tag{1.30}$$

Between these two cases it will have emission wings and an absorption core.

1.5. Flux Redistribution

If there are large scale motions within the envelope, then flux may be redistributed so that there are emission and absorption components to lines even though their total equivalent width is zero. For an expanding shell, Rottenberg (1952) has shown that emission and double emission components centred on the undisplaced wavelength, with a violet-displaced absorption component, may be formed (figure 1.4). By including fluorescence he obtains good fits to a variety of P Cygni type profiles.

1.6. Forbidden Lines

In very extended nebulae emission lines are found which are due to forbidden transitions of various atoms or ions. Before their identification as forbidden lines, they were attributed to a new element – nebulium – and one still occasionally finds them referred to in the literature as nebulium lines. They are mainly lines of [O II], [O III], [N I], [N II], [N III], [S II], [Ne III] and [Fe V].† Their upper levels are metastable (i.e. with only forbidden transitions to the ground and other lower states) with lifetimes of 10^{-1} to 10^{7} seconds, and with energies within a few electronvolts of the ground state. They are populated by collisional excitation from the ground state by electrons whose temperature is up to 20 000 K. Since the metastable level will be depopulated by collisional de-excitation or by further collisional or radiative excitation, both the particle and the radiation densities must be low for the production of forbidden emission lines. Hence they are mainly found in planetary nebulae or extended nova remnants, when the electron density is 10^{10} m^{-3} or less, and the dilution factor 10^{-10} or less.

† Spectroscopic Notation. The use of square brackets around a transition, e.g. [O III] 436.3, or [Ne V 342.6], or in the abbreviated form used above, denotes a transition which is forbidden under the normal electric dipole transition selection rules. It is, however, allowable under other (e.g. magnetic dipole, electric quadrupole etc) selection rules. But the transition probability will be typically 10^{-8} or less of a normal electric dipole transition probability. Lines resulting from such transitions are called forbidden lines.

A single bracket, e.g. Fe I 437.6] denotes an intercombination line. That is, a line from a transition between two terms of differing multiplicity. For example, the O I 135.6] line results from a transition between the two terms, ^3P–$^5S^0$. The transition probability for these lines is typically 10^{-6} that of a normal transition.

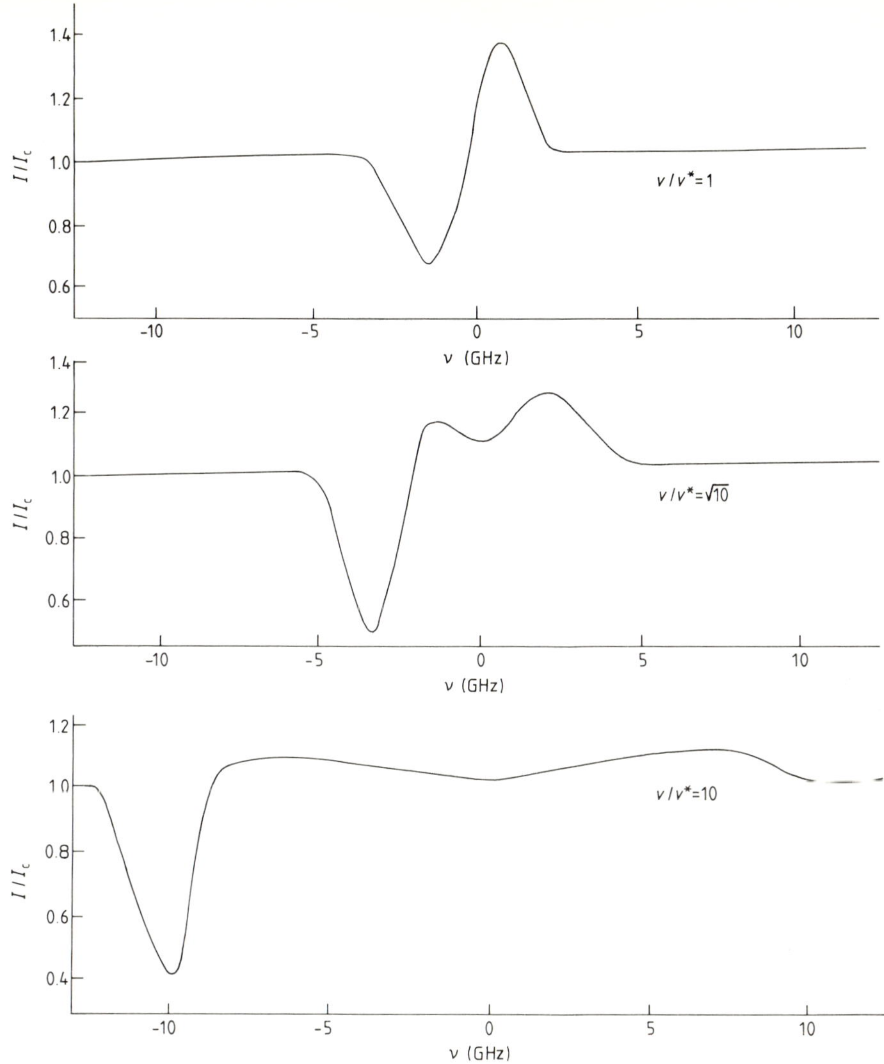

Figure 1.4. Line profiles formed by flux redistribution. v is the velocity of expansion and v^* is the most probable random velocity. (Reproduced from Rottenberg 1952 by permission.)

1.7. Line Profiles in Extended and Expanding Envelopes

Reasonably reliable line profiles in spherically symmetric moving envelopes can now be obtained, but immense amounts of time on the largest and fastest of computers are required. Little headway has been made with more complex (e.g. cylindrically symmetrical) cases. Work in this area is perforce therefore restricted to a comparatively few groups of workers by the computing power that is required. Some of their results which apply to particular models and which may be applied to some of the stars and star systems discussed later, are included in the appropriate chapters later. Here, only the main ideas behind the computations are considered. For more detailed treatments, the reader is referred to some of the recently available work and reviews (Castor and Lamers 1979, Drake and Ulrich 1980, Hummer 1976, Hummer and Rybicki 1967, 1971, Kalkofen 1970, Kogure *et al* 1978, Mihalas 1978, 1979, Poeckert and Marlborough 1978, Rybicki 1970, Surdej 1979, Underhill 1970, Wellman 1955).

Radiation in an envelope may be altered in four main ways:

(*a*) its direction may be changed,
(*b*) its frequency may be changed,
(*c*) energy may be added to the radiation field,
(*d*) energy may be subtracted from the radiation field.

The first of these can be important near the star where outwardly moving radiation which is coherently scattered (i.e. unchanged except for its direction of motion) may be re-absorbed by the star (or by the optically thick core of an envelope). Frequency non-coherence is an important concept and complete redistribution is often assumed (i.e. there is no correlation between the frequencies before and after scattering; both are independently distributed over the line profile). The most important process leading to non-coherence is the Doppler effect. Radiation in the line emitted by a moving atom will be 'seen' by a second atom (which in general will be moving at a different velocity) at a different frequency from that at which it was emitted. If this second atom then scatters the radiation, the photon will be at a third frequency which will depend on the projection of the velocity of the second atom on to the direction of the original photon, and so on. Natural and collisional broadening of the atomic levels also lead to changes in the frequency of the scattered photon, as does recoil of the atom during scattering.

Although not usually considered under this heading, much larger changes in frequency can occur. The Rosseland cycle which degrades high-energy radiation to lower energies has already been considered. Another process is absorption and re-emission by dust which converts ultraviolet and optical radiation into long-wavelength infrared and microwave radiation.

Emission and absorption in envelopes is largely the interchange of energy between thermal, excitation/ionisation and radiative states. The sources of radiation include collisional excitation, exchange of excitation energy between atoms, direct and dielectronic recombination with emissions occurring as the electrons cascade downwards. 'Sinks' of radiative energy include photo-ionisation with three-body recombination, collisional de-excitation and collisional ionisation from excited states.

One important concept which simplifies many problems is leakage or escape probability. First proposed by Sobolev (1953, 1960), it is simply that in an envelope with large velocities and a large velocity gradient, a line photon some distance from its point of origin will be Doppler shifted out of the line as 'seen' by an atom in its vicinity. The photon therefore has a far higher probability of escaping completely from the envelope than in the static case. In an expanding shell, a photon escaping at the inner edge of the shell is likely to escape completely, since, when it reaches the opposite side of the nebula, it will be Doppler shifted out of the line as seen by an observer moving with the shell material. In a thick shell there is a geometric localisation of interaction with the photon and the main concern of a theorist becomes the calculation of the escape probability of a photon from its interaction zone.

1.8. Stellar Winds

A recurring feature of the stars discussed in later chapters is their possession of an expanding atmosphere or envelope. The observational indications of this are manifold; P Cygni line profiles, asymmetric absorption lines (with extended short-wavelength wings), multiple absorption components to lines (with the components displaced to shorter wavelengths than the rest position of the line), lines displaced to shorter wavelengths whose displacement varies with excitation or ionisation, and so on. The expansion velocities which are indicated range from a few tens to many thousands of kilometres per second. In novae the expansion seems likely to be primarily due to a single violent explosion, possibly due to thermonuclear runaway at the surface of a white dwarf. In Be stars, the rapid rotation of the star may be involved, while in close binary systems, co-rotation with the binary will push material outwards. These mechanisms, however, cannot be invoked to explain mass loss in Wolf–Rayet stars, planetary nebulae, OB supergiants etc. Furthermore, in close binary systems and Be stars, additional processes to those indicated above must be involved if the material is to be lost completely, otherwise only expansion near to the star would result. Even in novae, material may continue to be ejected in a non-violent manner after the initial explosion, and so we require a supplementary ejection mechanism.

This (relatively) steady outflow of material is called the stellar wind. It is an extremely common phenomenon. A weak wind ($10^{-14} M_\odot$ a^{-1}) occurs in the Sun, and therefore presumably also in other solar type stars. The theory of the solar wind suggests that it will exist for any star with a hydrogen convection zone, i.e. stars of spectral type F or later, or 95% of all stars! Stronger winds are found in red giants and, as already seen, they occur in many varieties of early type stars. Two basic accelerating mechanisms for the material in a stellar wind have been suggested: thermal pressure and radiation pressure. The first of these leads to coronal models of the wind which successfully explain the main features of the solar wind. The second leads to a number of different models. In most of these the pressure is due to absorption in the ultraviolet by the resonance lines of the commoner elements, and they vary only in the initial accelerating mechanisms that they assume. In red giants, the pressure may be due to absorption by dust grains, which then drag the gas with them as they are pushed outwards. (The latter process depends crucially on whether or not the grains can condense out close enough to the star, and this in turn depends on the purity of the grains. The reality of this agency for a stellar wind has therefore yet to be firmly established theoretically.) There is also a hybrid model where a thin corona provides material with an initial acceleration, which is then further accelerated by ultraviolet radiation pressure.

1.8.1. Coronal Model

Parker (1958) realised that the high temperature (10^6 K) of the Solar Corona implied a coronal pressure that would never be reduced to zero, or even to the pressure of the interstellar medium (10^{-12} N m^{-2}), by the solar gravity. The corona cannot, therefore, be in hydrostatic equilibrium and must be in a state of constant expansion. The coronal number density scale height (H) is about 10^8 m, and its temperature ($T(r)$) is given approximately by

$$T(r) = T_0 r^{-2/7}, \tag{1.31}$$

where r is the distance from the centre of the Sun in solar radii and T_0 is the temperature at the base of the corona (1.5×10^6 K).

If the corona were in hydrostatic equilibrium, then the equation of hydrostatic equilibrium,

$$\frac{\mathrm{d}P(r)}{\mathrm{d}r} = -\frac{GM_{\odot}\rho(r)}{r^2}, \tag{1.32}$$

could be written:

$$\frac{\mathrm{d}(n(r)\,T(r))}{\mathrm{d}r} = \frac{-GM_{\odot}m_{\mathrm{H}}}{2k}\frac{n(r)}{r^2}, \tag{1.33}$$

(note that, since the coronal material is largely ionised hydrogen,

$$P(r) = 2n(r)\,kT(r) \qquad), \tag{1.34}$$

where $P(r)$ is the pressure at distance r, $\rho(r)$ is the density at distance r, $n(r)$ is the proton or electron number density at distance r and m_{H}, $M_{\odot}$, G and k have their usual meanings.

Substituting for $T(r)$ in equation (1.33) and using

$$H = \frac{2kT_0}{GM_{\odot}m_{\mathrm{H}}}, \tag{1.35}$$

we get:

$$\frac{\mathrm{d}(n(r)\,r^{-2/7})}{\mathrm{d}r} = -\frac{n(r)}{Hr^2}, \tag{1.36}$$

which on expanding gives:

$$\frac{\mathrm{d}n(r)}{\mathrm{d}r}=n(r)\left(\frac{2}{7r}-\frac{1}{Hr^{12/7}}\right). \tag{1.37}$$

Defining n_0 as the number density at $r = 1$, we have on integrating equation (1.37):

$$n(r)=n_0r^{2/7}\exp\left(\frac{7}{5H}(r^{-5/7}-1)\right). \tag{1.38}$$

Therefore, from equation (1.34), the pressure is given by

$$P(r)=P_0\exp\left(\frac{7}{5H}(r^{-5/7}-1)\right), \tag{1.39}$$

which for $r = \infty$ gives:

$$P(\infty)=P_0\exp\left(\frac{-7}{5H}\right). \tag{1.40}$$

For the Sun

$$P_0\simeq 2\times 10^{-2}\ \mathrm{N\ m^{-2}} \tag{1.41}$$

and

$$H\simeq 0.14R_{\odot}, \tag{1.42}$$

so that

$$P(\infty)\simeq 10^{-6}\ \mathrm{N\ m^{-2}}, \tag{1.43}$$

which is about a million times larger than the pressure of the interstellar medium.

The velocity function of the wind is critically dependent upon the initial conditions. For a steady state, spherically symmetric wind, we have:

$$\dot{M}=\text{constant}=4\pi r^2r_0^2\rho(r)\,v(r) \tag{1.44}$$

where $\dot{M}$ is the mass loss rate, r_0 is the distance to the base of the corona from the centre of the star in metres (normally essentially the same as the stellar radius) and $v(r)$ is the velocity at distance r.

The momentum balance equation is:

$$\rho(r)\,v(r)\frac{\mathrm{d}v(r)}{\mathrm{d}r}=-\frac{\mathrm{d}P(r)}{\mathrm{d}r}-\frac{\rho(r)\,GM_{\odot}}{r^2r_0} \tag{1.45}$$

which on substituting

$$P(r)=\frac{2\rho(r)\,kT(r)}{m_{\mathrm{H}}} \tag{1.46}$$

and assuming an isothermal corona:

$$T(r) = T_0, \tag{1.47}$$

becomes:

$$v(r)\left(1 - \frac{2kT_0}{m_H} v(r)^{-2}\right)\frac{dv(r)}{dr} = -\frac{GM_\odot}{r_0}\left(1 - \frac{4kT_0 r_0}{Gm_H M_\odot} r\right) r^{-2}. \tag{1.48}$$

For the Sun

$$\frac{4kT_0 r_0}{Gm_H M_\odot} \simeq 0.26, \tag{1.49}$$

so that for $r = 3.8R_\odot$ the right-hand side of equation (1.48) becomes zero. This is called the critical radius, r_c. On the left-hand side of equation (1.48) at $r = r_c$, one or more of the following possibilities must hold:

$$v(r_c) = 0 \tag{1.50}$$

$$v(r_c) = \left(\frac{2kT_0}{m_H}\right)^{1/2} \tag{1.51}$$

$$\frac{dv(r_c)}{dr} = 0. \tag{1.52}$$

For $r_0 < r < r_c$, the right-hand side of equation (1.48) is negative, and for $r > r_c$, it is positive. The velocity given by equation (1.51) is the isothermal sound velocity (v_c). The family of solutions of equation (1.48) is shown in figure 1.5. The condition given by equation (1.52) results in solutions of types 1, 2, 3 and 4. Type 1 winds are everywhere subsonic (sometimes called stellar breezes!). Types 3 and 4

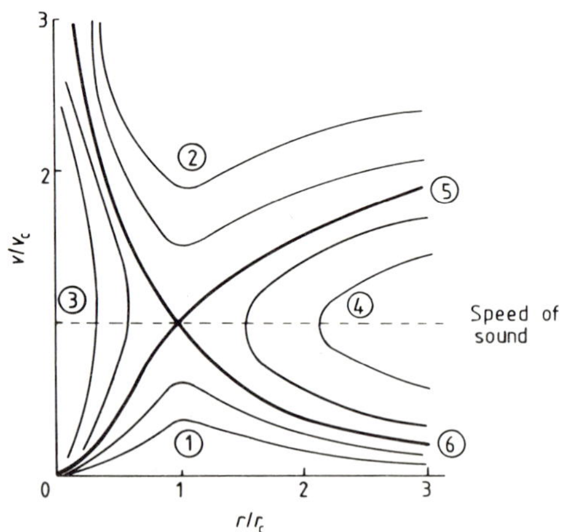

Figure 1.5. Stellar wind velocity functions for an isothermal coronal type wind. (From *Stellar Atmospheres* by D Mihalas. W H Freeman and Co. Copyright © 1978.)

predict two values of the velocity for any given distance from the Sun, and cannot therefore provide complete models (although type 4 winds can be used to provide part of the solution for models which include shock transitions to the interstellar medium). Type 5 and 6 solutions (shown by the bold curves in figure 1.5) occur when condition (1.51) is satisfied, and are particularly interesting as they are trans-sonic winds. They provide the only solutions which allow the velocity to change from subsonic to supersonic (or vice versa). The critical solution, type 5, is the one which models the solar wind and which is most useful when considering stellar winds. Condition (1.50) leads to $v(r) = 0$, and so does not produce a wind.

The isothermal assumption breaks down for large values of r, and equation (1.48) becomes, in the non-isothermal case,

$$v(r)\left(1-\frac{k}{\mu m_{\rm H}}\frac{T(r)}{v(r)^2}\right)\frac{\mathrm{d}v(r)}{\mathrm{d}r}=-\frac{GM_\odot}{r_0}\left(1-\frac{2kr_0}{\mu m_{\rm H}GM_\odot}rT(r)\right)\frac{1}{r^2}-\frac{k}{\mu m_{\rm H}}\frac{\mathrm{d}T(r)}{\mathrm{d}r}. \qquad (1.53)$$

The energy equation is then:

$$E=\frac{2\pi r_0^2}{\mu}\rho(r)\,v(r)^3r^2-\frac{4\pi GM_\odot r_0}{\mu}\rho(r)\,v(r)\,r+\frac{20\pi kr_0^2}{\mu m_{\rm H}}\rho(r)\,v(r)\,r^2T(r)$$

$$-4\pi\kappa_0 r_0 r^2\frac{\mathrm{d}T(r)}{\mathrm{d}r}, \qquad (1.54)$$

where E is the total energy flux in the stellar wind, μ is the mean particle weight in atomic mass units and κ_0 is the thermal conductivity at the reference level. A large number of solutions of equations (1.53) and (1.54) are given by Durney and Roberts (1971) for a wide range of initial parameters.

The basic source of energy for coronal winds is not entirely clear. It is suspected, however, that it is largely acoustic energy generated in the hydrogen convection zone, and propagated outwards into the corona via a variety of magnetohydrodynamic wave phenomena. Once in the corona, the wave energy is converted into thermal energy by the generation of shock waves through the 'whip-lash' effect as the coronal density decreases.

For early type stars, the observed rates of mass loss can be as high as $10^{-5}\,M_\odot\,\mathrm{a}^{-1}$. Coronal temperatures of over 10^8 K would be required to produce such a mass loss rate. Since such temperatures would produce other effects (e.g. soft x-ray emission, absence of lines due to low stages of ionisation etc) which are not found, different mechanisms must be sought for the winds in these stars. Such mechanisms are found in the radiatively driven models already mentioned. However, in these models, the material can be driven outwards very efficiently once it is already moving rapidly, but severe difficulties are encountered in accelerating the material from rest. Coronal winds may therefore play a part in early type stars as the initial accelerating mechanism of their winds. The final model is a hybrid with a very thin coronal stellar wind at the base of a much more extensive radiatively driven wind whose effects are those which are actually observed.

1.8.2. *Radiatively Driven Winds*

The basic accelerating mechanism of these winds is that of absorption of radially directed radiation followed by isotropic re-emission. Since a photon carries with it a momentum given by

$$p = h\nu/c, \tag{1.55}$$

all of which is directed outwards on absorption, but which is equally distributed in all directions on re-emission, there is a net outward force on the absorbing particles. Lucy and Solomon (1970) developed the first version of this model for absorption in the strong ultraviolet resonance lines. For the C IV line at 154.8 nm near the surface of an O type supergiant they found (assuming that the atom is absorbing unattenuated continuum radiation)

$$g_R = 2900(N_{\mathrm{C\,IV}}/N_{\mathrm{C}})\ \mathrm{m\ s^{-2}} \tag{1.56}$$

where g_R is the radiative acceleration, $N_{\mathrm{C\,IV}}$ is the number density of the C IV ions and N_{C} is the number density of all carbon atoms and ions. Since the gravitational acceleration is about 10 m s^{-2}, the ratio of radiative to gravitational acceleration is about 300 at maximum (when $N_{\mathrm{C\,IV}}/N_{\mathrm{C}} = 1$). This is an upper limit since normally the line flux will be very much reduced by absorption in the photosphere when compared with the nearby continuum flux. However, Lucy and Solomon showed that even allowing for photospheric absorption, radiative acceleration could still exceed gravitational acceleration. Once the material begins to move, the radiation field will be Doppler shifted with respect to an absorbing atom. The atoms will then 'see' radiation from the wings of the line or from the continuum, and the excess of radiation over gravitation will approach the limit given above. The accelerated atoms or ions will rapidly collide with other particles in the atmosphere and the momentum will be quickly shared out amongst all the material. Hence the atmosphere as a whole will experience an outward force.

Including radiation, the momentum equation becomes (Mihalas 1978):

$$v(r)\left(1 - \frac{k}{\mu m_{\mathrm{H}}}\frac{T(r)}{v(r)^2}\right)\frac{\mathrm{d}v(r)}{\mathrm{d}r} = \frac{2k}{\mu m_{\mathrm{H}}}\frac{T(r)}{r} - \frac{k}{\mu m_{\mathrm{H}}}\frac{\mathrm{d}T(r)}{\mathrm{d}r} - \frac{GM_*}{r_0}\frac{1-\Gamma(r)}{r^2} \tag{1.57}$$

where $\Gamma(r)$ is the ratio of radiative and gravitational accelerations. In the isothermal approximation this becomes

$$v(r)\left(1 - \frac{kT_0}{\mu m_{\mathrm{H}}}v(r)^{-2}\right)\frac{\mathrm{d}v(r)}{\mathrm{d}r} = -\frac{GM_*}{r_0}\left(1 - \Gamma(r) - \frac{2kT_0 r_0}{\mu m_{\mathrm{H}} GM_*}r\right)r^{-2}, \tag{1.58}$$

which is almost identical, except for the radiative term, $\Gamma(r)$, in the right-hand side, with equation (1.48). If we are to obtain solutions similar to the type 5 trans-sonic solution of equation (1.48), then for some value of r, r_c', the velocity, $v(r_c')$, must become equal to the velocity of sound, i.e.

$$v(r'_c) = v'_c = \left(\frac{kT}{\mu m_H}\right)^{1/2} \tag{1.59}$$

This value makes the left-hand side of equation (1.58) zero so that, from the right-hand side,

$$1 - \Gamma(r'_c) - \frac{2kT_0 r_0}{\mu m_H GM_*} r'_c = 0, \tag{1.60}$$

which has no solutions for

$$\Gamma(r'_c) > 1 - \frac{2kT_0 r_0}{\mu m_H GM_*} r'_c. \tag{1.61}$$

Thus for values of Γ *above* this critical value, the radiatively driven wind *cannot* pass from a subsonic to a supersonic regime. We thus have the paradoxical situation that if the radiative acceleration is greater than the fraction of the gravitational aceleration given by equation (1.61), no trans-sonic solutions exist and a subsonic wind will remain subsonic. The flow in such a case is compressed by the radiation pressure rather than accelerated (Marlborough and Roy 1970), and in fact the wind decelerates. To quote Hearn (1979):

> 'This results seems to be an excellent example of the sheer perversity of nature; the harder you push it, the slower it goes!'

Once the wind has passed the speed of sound, then radiation pressure accelerates it very efficiently, as outlined earlier. The terminal velocity can easily reach thousands of kilometres per second (which is the velocity indicated in Wolf–Rayet stars etc). The effect of subordinate lines is to increase the mass loss rate, in some cases by up to a factor of 100, without increasing the terminal velocity (Castor *et al* 1975).

The problem in a radiatively driven wind then, is to accelerate the material through the speed of sound. This initial accelerating mechanism is still the subject of dispute, and several hypotheses exist. It may well be that different mechanisms are important in different types of star, or act at different stages of the acceleration process. Two possible mechanisms have already been mentioned:

A: The radiation pressure is reduced near the star, and at low velocities, if there is a strong photospheric absorption line which reduces the flux. This may reduce Γ to a sufficiently low figure that the wind may become supersonic before the radiation field becomes Doppler shifted by enough to raise Γ above its critical value (Castor *et al* 1975). For this process to be possible, the photospheric line must be appreciably broader than the Doppler shift corresponding to v_c, i.e.

$$\frac{\Delta\lambda}{\lambda} > \left(\frac{2kT}{\mu m_H c^2}\right)^{1/2}. \tag{1.62}$$

B: A very thin coronal type wind may be present at the base of the radiatively driven wind and provide supersonic velocities via a trans-sonic solution of type 5 (Hearn 1979).

Other mechanisms which may operate include:

C: Stochastic instabilities in the radial velocity. Such a process has been proposed for Of stars (and may also apply to Wolf-Rayet stars) (Andriesse 1980) and it effectively leads to an additional pressure term in the momentum balance equation.
D: A fourth possibility was suggested by Rogerson and Lamers (1975) which also (like 'B') postulates a thin hot zone at the base of the stellar wind. In this region oxygen and nitrogen are in the form of O VI and N V, and provide a weak initial radial force. As the material expands, it cools and C III, C IV, N III and Si IV form and add to the radiation pressure. So again a trans-sonic solution is possible if the radiation pressure remains low until the velocity of sound is exceeded.

This final model also has the merit of explaining the observed presence of O VI and N V lines. The previous models must invoke shock waves or some other *ad hoc* process to produce local regions of high temperature and ionisation.

The rate of mass loss in a radiatively driven wind was shown by Lucy and Solomon (1970) to be independent of the wind velocity, and to be given by

$$\dot{M} \lesssim \frac{L_*}{c^2} N_{\rm L} \qquad \mathrm{kg\ s^{-1}} \tag{1.63}$$

$$\lesssim 7 \times 10^{-14} \left(\frac{L_*}{L_\odot}\right) N_{\rm L} \qquad M_\odot\ \mathrm{a^{-1}}, \tag{1.64}$$

where $N_{\rm L}$ is the number of strong ultraviolet resonance lines and L_* is the luminosity of the star. Allowing for the effect of the subordinate lines (Castor *et al* 1975)

$$\dot{M}_{\rm max} \leqslant \frac{L_*}{v_\infty c} \qquad \mathrm{kg\ s^{-1}},$$

where v_∞ is the final velocity of the stellar wind.

Non-spherically symmetric situations leading to stellar winds have been considered by Wolfson (1980), who studied winds initiated by accretion. He looked at two models:

(*a*) A single accretion stream, with outflow occurring on the opposite side of the accreting mass.

(*b*) Accretion in a disc with two outflowing streams perpendicular to the disc. (This situation may be reversed to give two accreting streams and outflow in a disc.)

The flow fields of his models are shown in figure 1.6. The second model may be applicable to SS433 (see chapter 6).

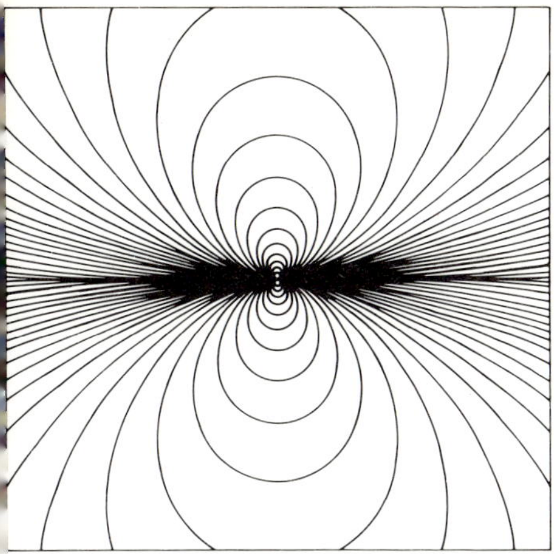
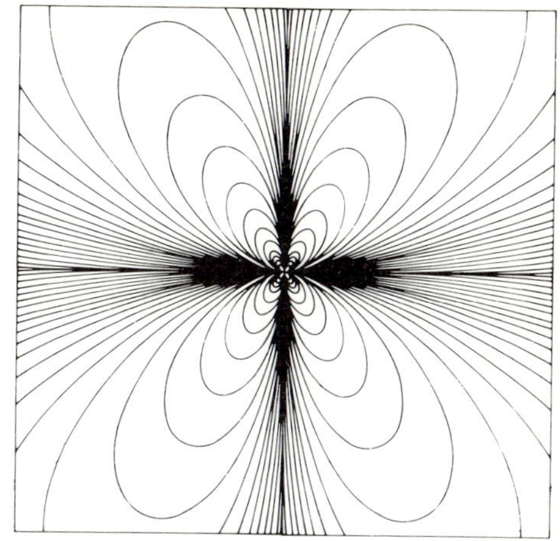

Figure 1.6. Flow fields for asymmetric stellar winds. Left: single accretion stream, and single ejection stream. Right: disc accretion with two ejection streams (or vice versa). The solutions are self-similar so that no scales are required. (Reproduced from Wolfson 1980 by permission.)

1.8.3. Significance of Stellar Winds

Stellar winds have effects on many aspects of astrophysics:

(*a*) They contribute material to the envelope of the star, and in some cases are the entire source of the material of the envelope.

(*b*) The higher mass loss rates may have a significant effect on the evolution of the star.

(*c*) Material is returned to the interstellar medium. This material may have been processed to some extent inside the star, and so will contribute to the evolution of the composition of the interstellar material.

(*d*) For a star embedded in a gas cloud, the wind may carve out a relatively gas-free hollow around the star. Possible examples are the Rossette Nebula and the paraboloid bubble in the Orion Nebula (Gull and Sofia 1979). The giant H II regions in the Magellanic Clouds may also be partly due to stellar winds.

Undoubtedly many other effects will become apparent as stellar winds become better understood and better observed over the next few years.

2. *Wolf-Rayet Stars*

2.1. Introduction

In 1867 Messieurs C Wolf and G Rayet (Wolf and Rayet 1867) notified the French Academy of Sciences of their discovery of three stars whose spectra showed permanent emission lines. Prior to this notification, only one such star was known – the Be star, γ Cas (see chapter 5). These three stars were of eighth magnitude and were in Cygnus. Their emission lines were very broad and dominated the green–blue part of the spectrum. Soon afterwards Respighi found that HD 68273 (γ^2 Vel, magnitude 1.7) had a similar spectrum, and such stars became known as Wolf-Rayet stars (HD 68273 is their brightest representative). Since then, the total number of known Wolf–Rayet stars has risen to about 250 (Smith 1968a, Westerlund and Smith 1964, Wray and Corso 1972), and since their emission lines are easy to pick out on objective prism survey plates, the catalogues of Wolf–Rayet stars are believed to be complete, at least down to magnitude 12 or 13. (Wolf–Rayet stars of eleventh magnitude or brighter are listed in table 2.4 at the end of this chapter.)

There are several sub-classes of the Wolf–Rayet stars. First there are three distinct types of object which exhibit Wolf–Rayet spectra:

(*a*) the less massive components of massive binary systems,
(*b*) single stars, often associated with ring nebulae and belonging to population I,
(*c*) nuclei of some planetary nebulae.

Secondly the spectra may be divided morphologically into two broad sub-classes:

(*a*) WN stars, in which emission lines of nitrogen and helium predominate,
(*b*) WC stars, in which emission lines of helium, carbon and oxygen predominate.

Each of these classes may then be further subdivided, from WN3 to WN8, and from WC5 to WC9.

The Wolf–Rayet stars have very strong stellar winds, and seem likely to have lost a large fraction of their original mass, and to be almost pure helium stars. They have been linked with the Of stars, and have been suggested as supernova precursors. They have been tentatively identified with both pre- and post-main sequence evolutionary stages of massive stars. Thus, as well as having considerable interest in their own right, they may have a much wider significance with regard to stellar evolution.

2.2. Spectral Classification

The main features of a Wolf–Rayet spectrum are (Thomas 1968, Underhill 1973):

(*a*) The spectrum consists almost wholly of emission lines superimposed on a continuous spectrum which resembles that of an O or B star.

(*b*) If absorption lines occur, then they are very few, and are satellites on the short-wavelength side of emission lines. If a general absorption spectrum is seen, then it is attributed to a companion, and the Wolf–Rayet star is regarded as a member of a binary.

(*c*) The emission lines are very broad, and the widths are not always the same for all lines in a spectrum. The widths correspond to velocities of hundreds to thousands of kilometres per second if due to Doppler broadening.

(*d*) Lines from a wide range of ionisation and excitation states are present in a single spectrum. The excitation temperatures are generally much higher than the colour temperature estimated from the continuum.

(*e*) Strong lines of helium are present in all spectra, but some (the WN stars) show strong lines of nitrogen, while others (the WC stars) show strong lines of carbon and oxygen.

Sample spectra are shown in figure 2.1.

There are several systems for the detailed spectral classification of Wolf–Rayet stars (Beals 1938, Hiltner and Schild 1966, Walborn 1974) which are broadly similar. The most comprehensive is that due to Smith (Smith and Aller 1971) and

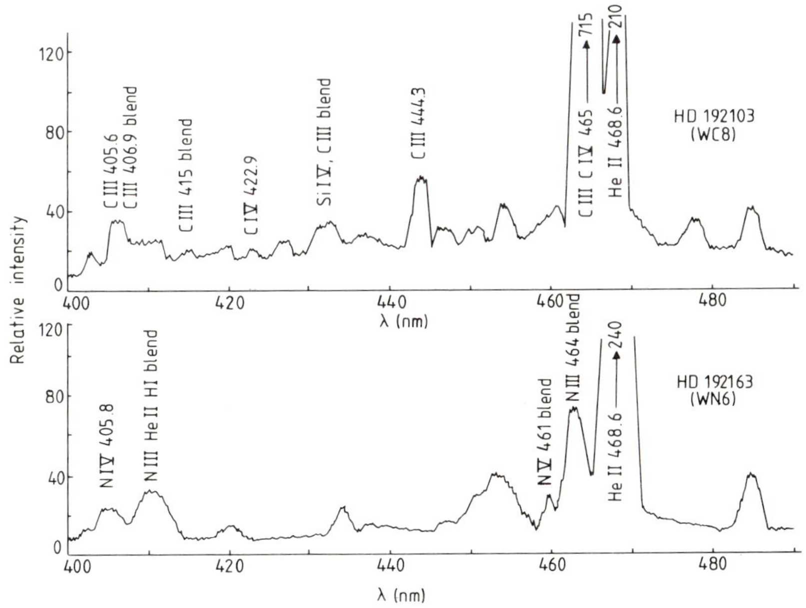

Figure 2.1. Blue spectra of HD 192103 (WC8) and HD 192163 (WN6).

she uses the following criteria:

WN Spectra. The classification depends on which ionisation state of nitrogen predominates. The lines used are:

N III: The blend from 463.4 to 464.1 nm and the line at 531.4 nm.
N IV: The blend from 347.9 to 348.4 nm and the line at 405.8 nm.
N V: The lines at 460.3 and 461.9 nm and the blend from 493.3 to 494.4 nm.

Helium lines are also used to separate WN7 and WN8 (see figure 2.2).

Class	Line ratios	
WN3	N IV $\ll$ N V	N III absent
WN4	N IV $\simeq$ N V	N III weak or absent
WN4.5	N IV > N V	N III weak or absent
WN5	N III $\simeq$ N IV $\simeq$ N V	N III 464 blend present
WN6	N III $\simeq$ N IV	N V weak, N III 464 blend present
WN7	N III $\gg$ N IV	He I weak, N III 464.0 < He II 468.6
WN8	N III $\gg$ N IV	N III 464.0 $\simeq$ He II 468.6, N III 531.4 present, He I strong with violet absorption edges

WC Spectra. The classification is based on lines of C III, C IV and O V as follows (see figure 2.3):

	Line ratios		
Class	C III 569.6 / O V 559.2	C III 569.6 / C IV 580.5	Width C III, IV 465.0 (nm)
WC5	<1.0	0.3	8.5
WC6	>1.0	0.3	4.5
WC7	8.0	0.7	3.5
WC8		1.0	
WC9		3.0	1.0

Stars in which an absorption spectrum is observed in addition to the Wolf-Rayet emission spectrum, or in which the strengths of all the emission lines are less than normal, are given a composite classification, WR + OB, and are taken to be binary systems. (Although at least one star, HD 193077 (WN5) seems unlikely to be a binary, and yet has highly rotationally broadened absorption lines of hydrogen and helium (Massey 1980). This star, however, has a hydrogen to helium ratio near unity and may be a newly formed Wolf-Rayet star, and so not typical.) Generally

Intensity
NIV 405.8
N III
He II
H I
blend
N III 464
blend
He II
468.6
WN6
WN7
WN8

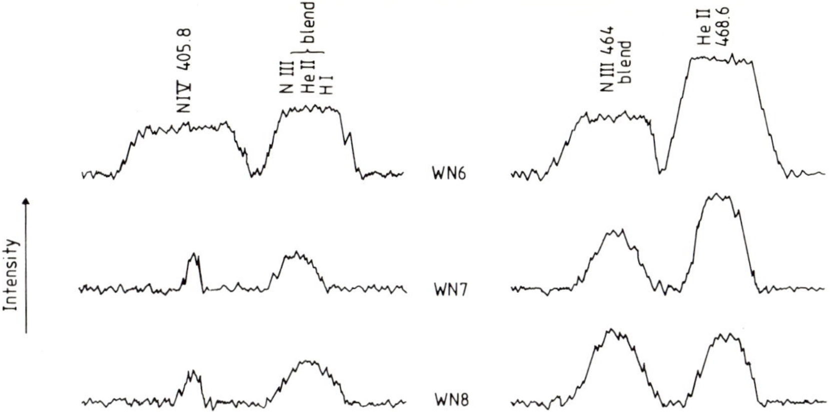

Figure 2.2. Emission line variation in WN stars.

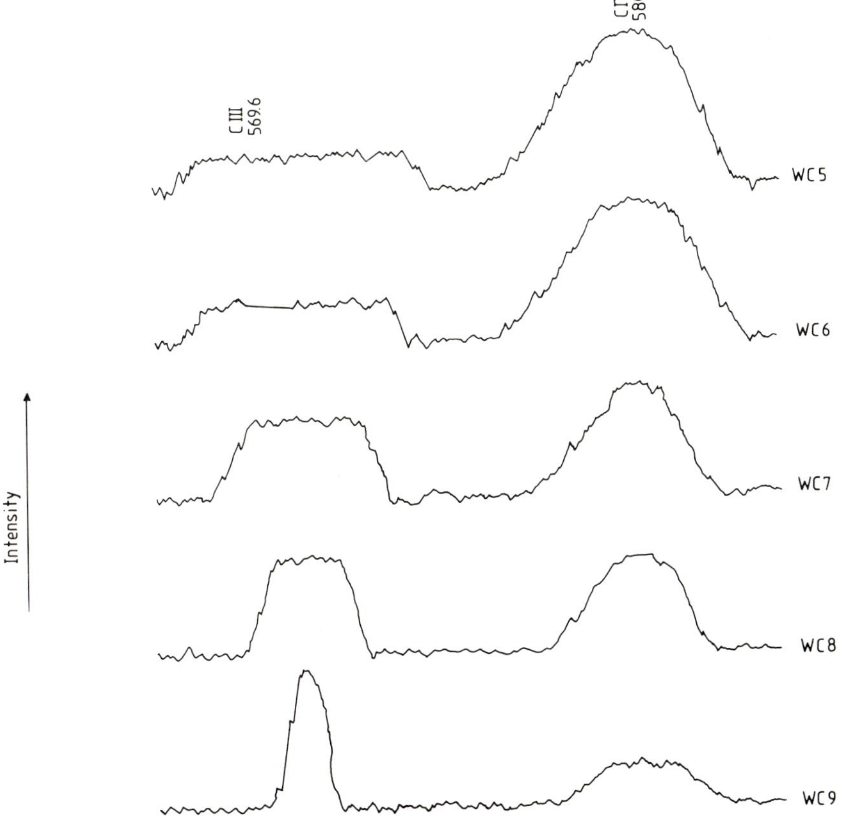

Intensity
C III
569.6
C IV
580.5
WC5
WC6
WC7
WC8
WC9

Figure 2.3. Emission line variation in WC stars.

there is little or no sign of hydrogen in Wolf–Rayet spectra, however, the later WN stars do show Balmer emission, and it appears that they may be related to the Of stars (see later in this chapter). Other lines due to C II, O II, O III, O IV, O VI, Si III and Si IV may be identified in some Wolf–Rayet spectra (Smith and Aller 1971). Members of the subset of very high excitation Wolf–Rayet stars, known as the O VI sequence (Smith and Aller 1969) have the O VI 381.1, 383.4 doublet as one of their strongest lines. They also contain lines due to O VII, O VIII and C V (Barlow *et al* 1980). Most (but not all) seem to be the central stars of planetary nebulae.

The emission lines are of two main types, flat-topped and rounded, as may be seen in figures 2.1, 2.2 and 2.3. Beals (1929, 1930, 1931, 1934) showed that such profiles could result for optically thin lines from atmospheres expanding at constant and uniformly accelerating velocities respectively: Consider a point, P, in the expanding atmosphere around a star, with coordinates (r, θ, ϕ) which are centred on the star and directed towards the observer (figure 2.4), and which has a velocity radial to the star of v. Then the velocity along the line of sight (the radial velocity in normal astronomical terminology) will be:

$$V_R = v \cos \theta \qquad (2.1)$$

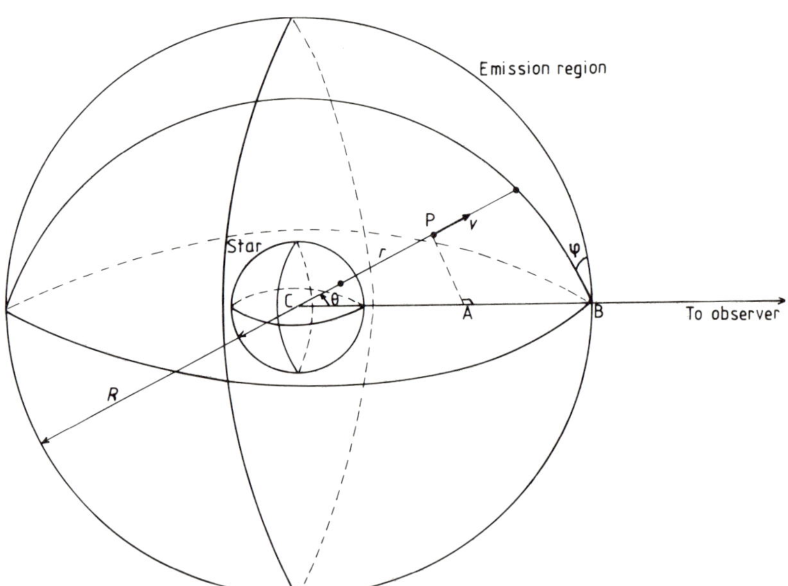

Figure 2.4. Spherical polar coordinate system.

Now consider two cases:

(*a*) constant velocity, i.e. $v = V$,

(*b*) uniform acceleration, i.e. $v = Vr/R$, where V is the maximum velocity attained and R is the radius of the emission region.

Ignoring occultation effects in both cases, we have for the first case

$$V_R = V \cos\theta \tag{2.2}$$

and for the second case

$$V_R = (V \cos\theta)\, r/R \tag{2.3}$$

$$= Vz/R \tag{2.4}$$

where

$$z = \text{CA} = r \cos\theta. \tag{2.5}$$

The constant velocity surfaces are thus cones centred on CB with their vertices at C in the first case, and planes perpendicular to CB in the second case. Now consider a segment of an emission line, at $\Delta\lambda$ from the line centre, width $d\Delta\lambda$ and height $I_{\Delta\lambda}$ (see figure 2.5). Then, ignoring other broadening mechanisms, the energy in the strip originates in the constant velocity zone which has velocity,

$$V_R = c\Delta\lambda/\lambda_c. \tag{2.6}$$

In the first case, the volume of the constant velocity zone between θ and $\theta + d\theta$ is given by

$$\text{Volume} = \int_0^R 2\pi r^2 \sin\theta \; d\theta \; dr \tag{2.7}$$

$$= \tfrac{2}{3}\pi R^3 \; d \cos\theta. \tag{2.8}$$

But

$$V \cos\theta = c\Delta\lambda/\lambda_c, \tag{2.9}$$

giving

$$d \cos\theta = \frac{c}{V\lambda_c}\, d\Delta\lambda \tag{2.10}$$

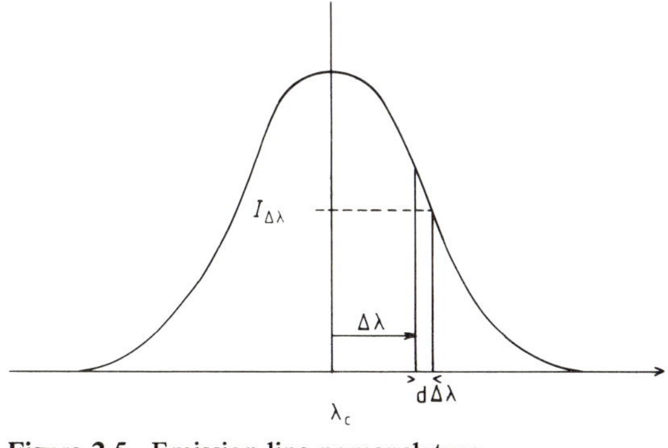

Figure 2.5. Emission line nomenclature.

and so,

$$\text{Volume} = \frac{2\pi c}{3V\lambda_c} R^3 \, d\Delta\lambda. \tag{2.11}$$

In the second case, the volume between z and $z + dz$ is:

$$\text{Volume} = \pi(R^2 - z^2)\, dz. \tag{2.12}$$

But

$$Vz/R = c\Delta\lambda/\lambda_c, \tag{2.13}$$

giving

$$dz = \frac{Rc}{V\lambda_c} d\Delta\lambda \tag{2.14}$$

and so,

$$\text{Volume} = \pi \left(R^2 - \frac{R^2 c^2}{V^2 \lambda_c^2} \Delta\lambda^2\right) \frac{Rc}{V\lambda_c} d\Delta\lambda. \tag{2.15}$$

Now the energy in the line segment is $I_{\Delta\lambda}\, d\Delta\lambda$. So, assuming that the emission per unit volume is constant over the atmosphere, we have:

$$I_{\Delta\lambda} \propto \begin{cases} \dfrac{2\pi c}{3V\lambda_c} R^3 & \text{in the first case} \quad (2.16) \\[2ex] \dfrac{\pi R^3 c}{V\lambda_c} \left(1 - \dfrac{c^2}{V^2\lambda_c^2} \Delta\lambda^2\right) & \text{in the second case.} \quad (2.17) \end{cases}$$

Thus the first case (optically thin, constant expansion) gives $I_{\Delta\lambda} \propto$ constant, i.e. a flat-topped line, while the second case (optically thin, uniform acceleration) gives $I_{\Delta\lambda} \propto (1 - \text{constant} \times \Delta\lambda^2)$, i.e. a parabola, or rounded-topped line.

Since both types of line profiles are observable in a single star, it is clear that at least one of Beal's assumptions must be wrong. Castor (1970) has undertaken a more rigorous study using Sobolev's escape probability method (see chapter 1). The escape probability, β, is the fractional excess of transitions from the ith state to the kth state, over the inverse reaction. Castor finds:

$$\beta = \int_0^1 \left(\frac{1 - \exp\{-\tau_0(r)[1 + \mu^2(d \ln V/d \ln r - 1)]^{-1}\}}{\tau_0(r)[1 + \mu^2(d \ln V/d \ln r - 1)]^{-1}} \right) d\mu \tag{2.18}$$

where

$$\tau_0(r) = \frac{\pi e^2}{mc} (gf)_{lu} \frac{(N_l/g_l - N_u/g_u)}{\nu V(r)/c} r \tag{2.19}$$

and the other symbols have their usual meanings. For a two-level atom he obtains the following expression for the line profile:

$$\frac{F_\nu - F_c}{F_c} = \frac{1}{r_c^2}\int_{r_{\min(\nu)}}^{\infty} \tau_0(r)\,\frac{S(r)}{I_c}\,\frac{1-e^{(-\tau)}}{\tau}\,2r\,dr$$

$$-\frac{1}{r_c^2}\int_0^{r_c} 1-\exp\{-\tau y[-(r_c^2-p^2)^{1/2}]\}\,2p\,dp$$

$$-\frac{1}{r_c^2}\int_0^{r} \frac{S[(p^2+z_0^2)^{1/2}]}{I_c}\,[\![\exp\{-\tau y[-(r_c^2-p^2)^{1/2}]\}$$

$$-\exp(-\tau)]\!]\,2p\,dp \tag{2.20}$$

where τ is a function of ν, p taken at the limit as $z \to \infty$ and is given by

$$\tau = \tau_0(r)\left[1 + \frac{z^2}{r^2}\left(\frac{d\ln V}{d\ln r} - 1\right)\right]^{-1}, \tag{2.21}$$

S is the source function given by

$$S = (1-\epsilon)\,\beta W I_c + \epsilon\beta_\nu, \tag{2.22}$$

ϵ is the proportion of collisional to total de-excitations of the upper level, W is the dilution factor, z is the coordinate along the axis through the star and the observer (increasing away from the observer), p is the radial distance away from the z axis, r is the radial distance away from the centre of the star, subscript c denotes the core (i.e. the star), y is given by

$$y[x] = \int_{-\infty}^{x} \phi(x')\,dx' \qquad \text{for } y(-\infty)=0, \quad y(\infty)=1, \tag{2.23}$$

and $\phi(\Delta\nu)$ is the profile function in the rest frame, integrated over frequency and normalised. A range of line profiles result (see figure 2.6) which have a single velocity law throughout the atmosphere (the law chosen by Castor to be

$$v = \left(1 - \frac{r_c}{r}\right)^{1/2} V_\infty \tag{2.24}$$

but without any intrinsic significance). The flat-topped profiles result from optically thin emission over a hollow sphere, and the rounded profiles occur for optically thick lines. Hence the assumption that all lines are optically thin is where Beals' analysis breaks down. Castor showed that lines are optically thin when their central intensity is less than 30% above the continuum.

The band widths of the emission bands are related to the spectral type (Beals 1940), decreasing from widths corresponding to velocities of 1600 to 2400 km s^{-1} at WC6 to velocities of 400 to 900 km s^{-1} at WC8, and from 600 to 2000 km s^{-1} at WN5 to less than 700 km s^{-1} at WN8.

Three WC9 stars have an excess optical continuum intensity compared with their emission line intensities. Although no general absorption line system occurs,

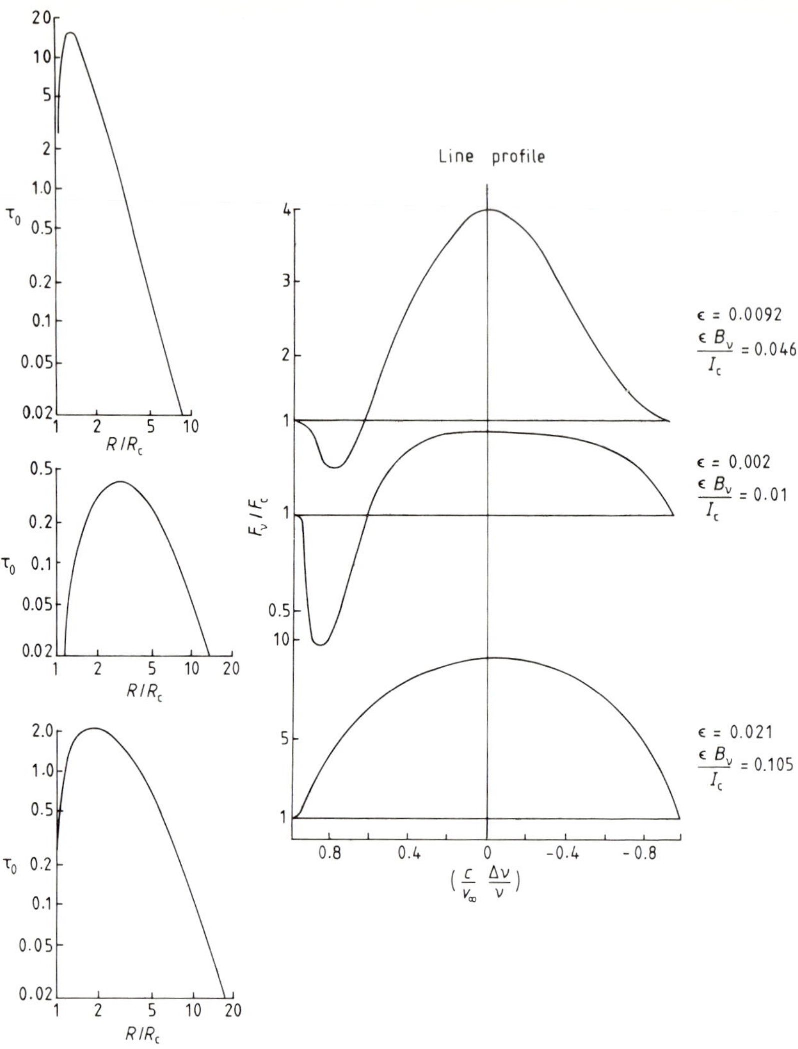

Figure 2.6. Wolf–Rayet type line profiles. (Reproduced from Castor 1970 by permission.)

the excess is attributed to the Wolf–Rayet star having a hot companion (Cohen and Kuhi 1977). Infrared excesses have been found in most Wolf–Rayet stars. For early WC stars and the WN stars this is attributed to free–free emission. The polarisation in HD 50896 (WN5) is attributed to scattering by free electrons (McLean *et al* 1979b) which lends support to this hypothesis. For the later WC stars, however, the excess seems to be due to thermal emission by circumstellar dust (Cohen 1975, Cohen *et al* 1975, Cohen and Vogel 1978, Gerhz and Hackwell 1974). A new dust shell formed around the WC7 + O5 system, HD 193793, between the middle of 1976 and the middle of 1977 (Williams *et al* 1978) leading to a sudden increase in its infrared excess. When first observed the new shell had a temperature of 900 K, and this then slowly cooled to 810 K in May 1978, 780 K in August and 650 K by October 1978 (Hackwell *et al* 1979, Williams and Antonopoulou 1979). The mass of the shell is variously estimated between 4×10^{-8} and $1.5 \times 10^{-4} M_{\odot}$ (Hackwell *et al* 1979), Williams *et al* 1978). The extreme WC9 system, Ve 2–45 suffers from an extinction of about 4.7^{m} which is also attributed to a surrounding dust shell (Cohen and Kuhi 1977).

In the near infrared spectra of WC stars (0.7 to 2.0 μm) numerous emission features due to H, He I, He II, C II, C III and C IV are found (Anglo-Austrian Observatory 1980a, Barnes *et al* 1974, Bernat *et al* 1977). There are also unidentified features at 1.72 and 2.41 μm (Williams *et al* 1980) (figure 2.7). At longer

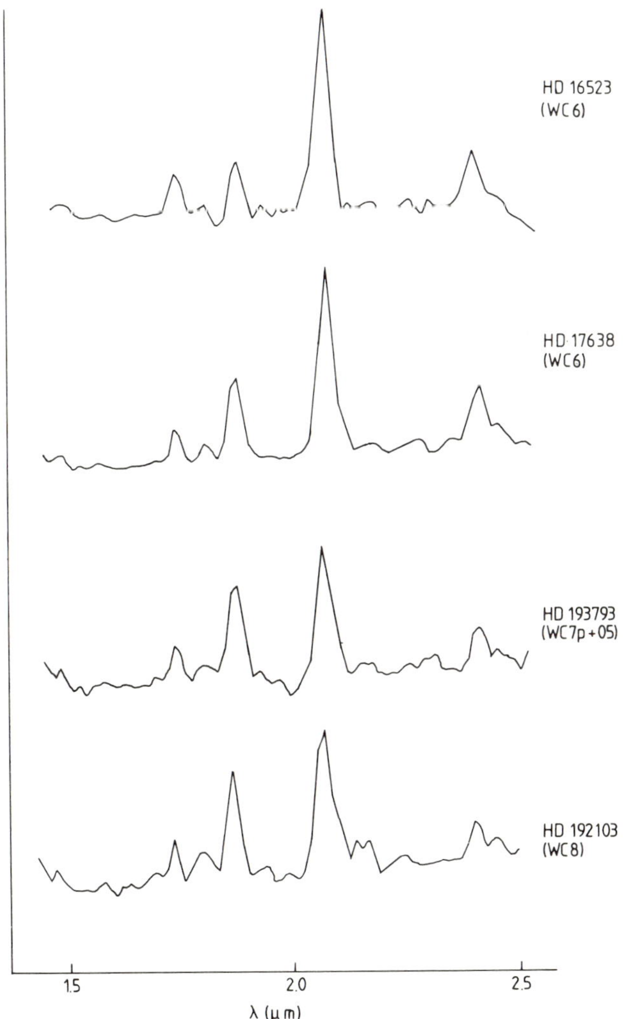

Figure 2.7. Near-infrared specra of WC stars. (Reproduced from Williams *et al* 1980 by permission.)

wavelengths there may be He I and He II features at 9.09, 10.48, 11.28 and 12.39 μm (Anglo-Austrian Observatory 1980b). A forbidden emission line of Ne II occurs at 12.81 μm (figure 2.8), and there may be unidentified features at 8.6 and 11.25 μm (Aitken *et al* 1980). The spectra of the later WC stars resemble those of planetary nebulae (see chapter 3). Violet-displaced absorption components for the He I, He II, N III, N IV, N V and C V emissions are found in HD 151932 (WN7) whose displacements increase with decreasing excitation potential (Seggewis 1974).

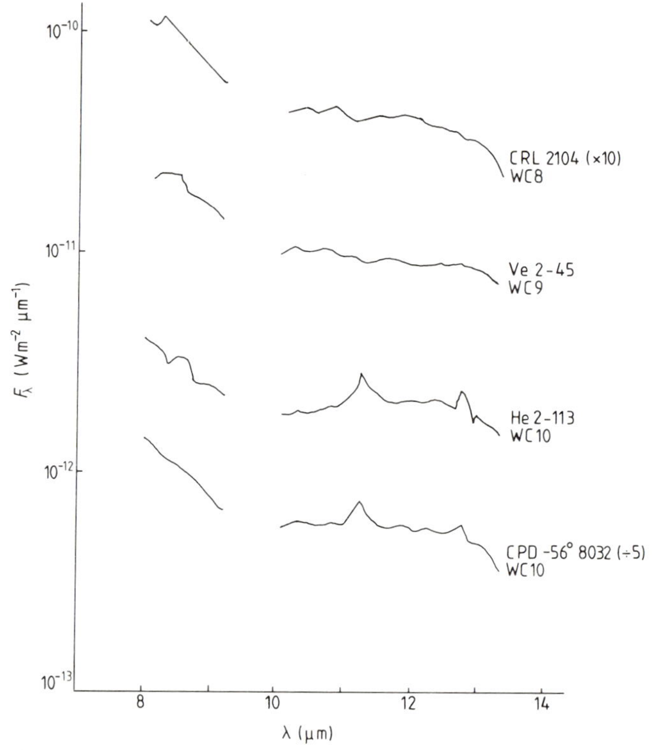

Figure 2.8. Infrared spectra of late WC stars. (Reproduced from Aitken *et al* 1980 by permission.)

The ultraviolet intensities vary on timescales ranging from minutes to months for both binary and single Wolf–Rayet stars (Burton *et al* 1978, Willis and Wilson 1976). In WN stars the principal ultraviolet lines are He II 164.0, N III 180.5, N IV 148.8, N IV 171.8 and C IV 154.8, while in WC stars they are C IV 154.8, 253.0, C III 190.9, C III 229.7, He II 164.0, N IV 148.8, N IV 171.8, Si IV 139.3 and 140.3 (Henize *et al* 1975, Johnson 1978, Willis and Wilson 1976) (see figure 2.9). The Si IV and C IV lines in HD 192163 (WN6) have two components. One set is undisplaced, and is thought to originate in the star. The other set is displaced by about 90 km s^{-1} to shorter wavelengths, and is thought to have been formed in

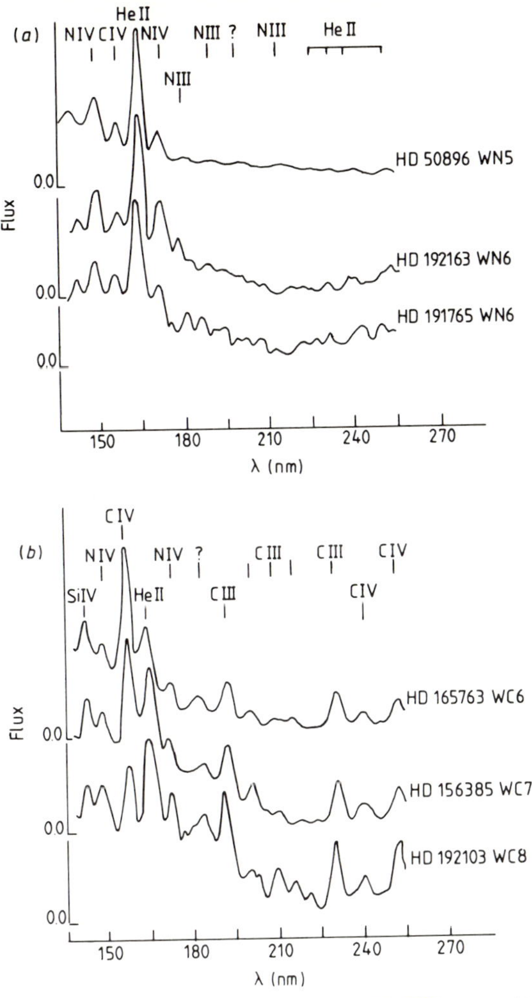

Figure 2.9. Ultraviolet spectra of (*a*) WN and (*b*) WC stars. (Reproduced from Willis and Wilson 1978 by permission.)

the expanding nebula, NGC 6888, around the star (Huber *et al* 1979). An anomalous excess at 220 nm has been observed in at least two stars, one of which is a WN6 and the other a WC7. It may be attributable to graphite in the material around the stars (Willis and Wilson 1977).

Several Wolf–Rayet stars have been detected at radio wavelengths. For example γ^2 Vel has been observed from 5 to 15 GHz, with a flux up to 67 ± 5 milli-Janskys (1 Jy $= 10^{-26}$ W m^{-2} Hz^{-1}) at the shorter wavelengths (Morton and Wright 1978, Seaquist 1976). HD 193793 emits 21 ± 1 mJy at 8 GHz and 26.2 mJy at 2.7 GHz (Florkowski and Gottesman 1977), while ζ Pup is just detectable at about 7 mJy at 15 GHz.

2.3. Luminosity and Temperature

The luminosity of most Wolf–Rayet stars is not well known, since their distances are only poorly determined via kinematic methods or via spectroscopic parallaxes. Further, the effect of the emission lines on the normal magnitude ranges may amount to 0.2^m. (Smith (1968b) has defined a system of narrow-band filters which avoids the emission lines in WN stars.) However, for those Wolf–Rayet stars in the Large Magellanic Cloud and in the cluster associations of Carina, Sco OB1 and TR 27, distances are reasonably well known and good estimates of the absolute visual magnitude, M_V, may be obtained. The data for single Wolf–Rayet stars are shown in figure 2.10 and although the number of stars is small, it appears that the WC stars all have absolute visual magnitudes near -4^m, the early WN stars have similar magnitudes, but the later ones are about 2^m brighter. The Wolf–Rayet stars which are the nuclei of planetary nebulae are much fainter, with absolute magnitudes in the range $+4^m$ to 0^m (Hack and Struve 1971).

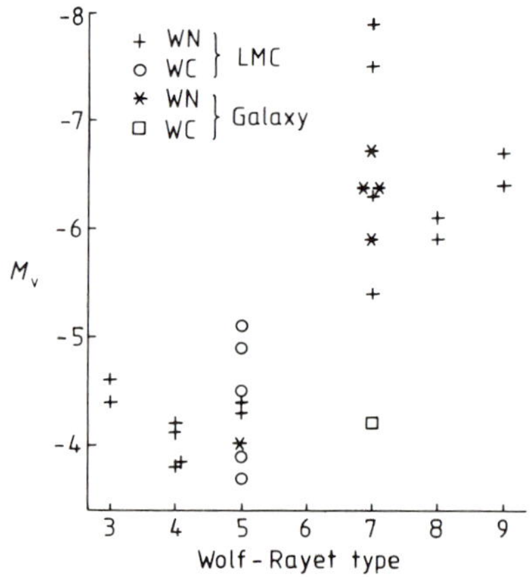

Figure 2.10. Absolute visual magnitude of single Wolf–Rayet stars. (Reproduced from Conti 1979 by permission.)

The temperature is more problematical. The earlier Wolf–Rayet stars seem, on the basis of several different methods of measurement, to have effective temperatures near 30 000 K, while the late WN stars show similarities with the Of stars, which indicates a temperature between 45 000 and 55 000 K, but have Zanstra temperatures of less than 25 000 K. The early WN stars have Zanstra temperatures of over 50 000 K. Zanstra temperatures using the He II 164.0 line and ultraviolet continuum flux distributions lead to temperatures near 30 000 K for both the WN5–6 and WC6–8 stars (Brune *et al* 1979, Willis and Wilson 1978). (A detailed discussion of the Zanstra method of temperature determination is given in chapter 3.)

The nuclei of planetary nebulae are again somewhat different, most being assigned temperatures near 100 000 K, with only a small proportion having temperatures as low as 30 000 K (Hack and Struve 1971).

These uncertainties in the temperature produce corresponding uncertainties in the bolometric magnitudes. The uncertainty is exacerbated by the probability that significant (perhaps 30%) amounts of energy are lost as mechanical energy in the stellar winds of Wolf-Rayet stars. Within these limits, the bolometric magnitudes of all Wolf-Rayet stars appear to be similar and around -9^m (see figure 2.11) but with a range of $\pm 2^m$.

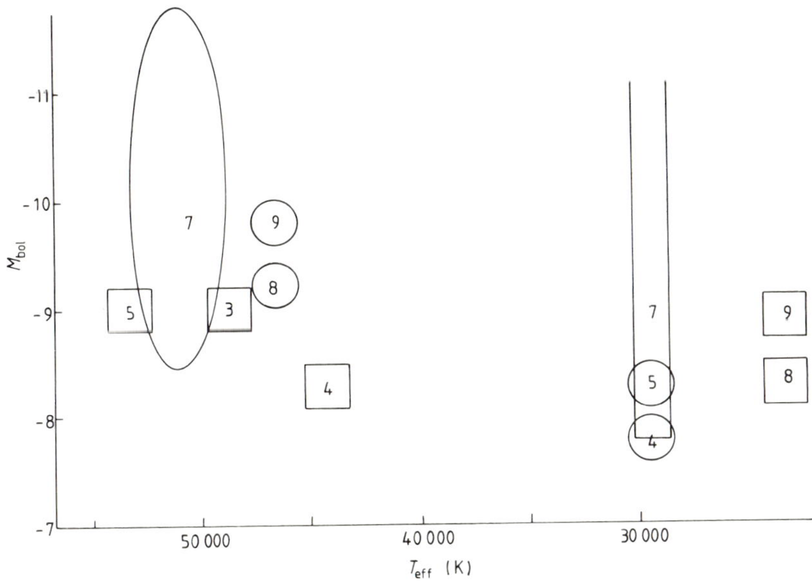

Figure 2.11. M_{bol} and T_{eff} for Wolf-Rayet stars. The rectangular boxes indicate Zanstra temperatures, elliptical boxes temperatures obtained by other means (Conti 1979). The numbers indicate the Wolf-Rayet classes.

2.4. Radii

The luminosity of a star is proportional to R^2T^4 (neglecting mechanical energy loss), hence the problems outlined above with regard to the determination of M_{bol} and T_{eff} combine to render radius estimation highly indeterminate. For $M_{bol} \simeq -9^m$ and $T_{eff} \simeq 30\,000$ K, we have a radius of about 25 $R_\odot$ for all Wolf-Rayet stars. However, if the Zanstra temperatures are taken, then the early Wolf-Rayet stars have radii of about 8 $R_\odot$ and the later ones radii of about 40 $R_\odot$.

Studies of spectroscopic binary systems, which include Wolf-Rayet stars, indicate separations of 40 to 80 $R_\odot$ and the sum of the photospheric radii of the stars as less than 15 $R_\odot$. Studies of HD 68273 and HD 193576 (γ^2 Vel and V444 Cyg) give radii of the Wolf-Rayet components of about 7 $R_\odot$ (Sahade and Wood 1978). However, the use of the stellar interferometer suggests a radius for the Wolf-Rayet component of HD 68273 of 17 ± 3 $R_\odot$, and a semi-major axis of 425 $R_\odot$ (Hanbury-Brown *et al* 1970).

The size of the region producing the emission lines may be up to four times the size of that producing the continuum, and there may be an even larger electron-scattering region. The radio emission from HD 193576, if due to free–free emission, suggests a surrounding gas cloud with a radius of 10^{13} m (14 000 $R_{\odot}$) (Seaquist 1976). Many Wolf–Rayet stars are embedded in a visible nebula. Such stars are mostly of the WN sub-class, and their nebulae range from 2 to 18 pc across. One WC star, HD 92809 (WC6), however, is at the centre of a circular nebula some 6 to 10 pc in diameter (Lortet *et al* 1980), showing that the nebulae are not solely the prerogative of WN stars. The nebulae associated with Wolf–Rayet stars in the Large Magellanic Cloud are significantly larger than those associated with the Galactic Wolf–Rayet stars, ranging up to 200 pc in diameter (Chu and Lasker 1980). The reason for the difference is not clear, and it may just reflect observational selection.

2.5. Masses

Three eclipsing binary systems which contain Wolf–Rayet stars have been analysed (Conti 1979) giving masses:

Star	Type	Mass ($M_{\odot}$)
HD 168206	WC8	8.1
HD 193576	WN5	8.4
HD 211853	WN6	11.5

HD 68273 has minimum masses of 15 to 17 $M_{\odot}$ and 32 to 53 $M_{\odot}$ for its WC8 and O9I stars respectively (Moffat 1977, Niemela and Sahade 1980). Five other Wolf–Rayet stars which are members of spectroscopic binary systems have been studied and have an average mass ratio of 0.4. The primary in each case is an OB star whose mass may be expected to lie in the region of 10 to 30 $M_{\odot}$, so that the masses of the Wolf–Rayet stars in these cases probably lie in the range 4 to 12 $M_{\odot}$.

Smith and Paczynski (Paczynski 1973, Smith 1973) argue on theoretical evolutionary grounds that the masses should lie in the range 8.5 to 14 $M_{\odot}$. However, the nuclei of planetary nebulae which may show Wolf–Rayet characteristics are thought to have masses of only about 1 $M_{\odot}$.

Hence the single population I Wolf–Rayet stars and those which are the lighter components of massive binary systems seem to form one group, with masses near 10 $M_{\odot}$, while the nuclei of planetary nebulae form a second group of Wolf–Rayet stars, with mass about 1 $M_{\odot}$. Although in most binaries the Wolf–Rayet star is the lighter component, in HD 50896, a mass of 10 $M_{\odot}$ is suggested for the Wolf–Rayet star with a companion of mass 1.3 $M_{\odot}$ (McLean 1980). This suggestion, however, remains to be firmly established.

2.6. Composition

The outstanding features of the composition of Wolf–Rayet stars will already have become apparent; the lack of hydrogen, and the predominance of helium with nitrogen or carbon and oxygen. (A similar dichotomy has been found for some absorption-line OB stars. OBN stars have enhanced nitrogen and depleted carbon, the opposite occurs in OBC stars (Walborn 1976). The relationship of these stars to the Wolf–Rayet stars (if any) is not clear.) For some time it appeared that the difference between WN and WC stars was one of temperature, and not composition. However, ultraviolet observations and the improvement of techniques for modelling line transfer in expanding atmospheres have enabled better abundance determinations to be made. The carbon/nitrogen ratio now appears to be about a thousand times larger in WC stars than in WN stars (see table 2.1).

Table 2.1. Abundances of helium, carbon and nitrogen in Wolf–Rayet stars (Willis and Wilson 1978, 1979).

Star (HD)	Type	N_{He} (m^{-3})	N_C/N_{He}	N_N/N_{He}	N_C/N_N
50896	WN5	2.0×10^{17}	9.0×10^{-5}	1.7×10^{-2}	5.4×10^{-3}
192163	WN6	2.0×10^{17}	9.0×10^{-5}	3.5×10^{-2}	2.6×10^{-3}
191765	WN6	2.0×10^{17}	9.0×10^{-5}	1.9×10^{-2}	4.7×10^{-3}
192103	WC8	1.0×10^{17}	9.0×10^{-3}	3.2×10^{-3}	2.9

The abundance of hydrogen is very much less than the cosmic abundance in all Wolf–Rayet stars. In WC stars blending of hydrogen lines with carbon lines makes the abundance extremely uncertain; in fact it is possible to fit any abundance from the cosmic abundance to zero! However, the upper limit seems very unlikely, and it is reasonable to assign a figure much nearer the lower limit for these stars. In WN stars, an abundance may be determined. This is very low for the early stars, but increases, while still being less than the cosmic abundance, on progressing to later stars in the sequence (see table 2.2).

Table 2.2. Relative abundances of hydrogen and helium in Wolf–Rayet stars (Smith 1973).

Star (HD)	Type	N_H/N_{He}
Cosmic		10.0 ± 0.1
9974	WN3	0.8 ± 0.4
187282	WN4	0.4 ± 0.3
190918	WN4	0.0 ± 0.1
50896	WN5	0.0 ± 0.05
192163	WN6	$\lesssim 0.4 \pm 0.2$
211853	WN6	$\lesssim 0.3 \pm 0.2$
151932	WN7	<1.0
MR 119	WN8	$\ll 2.3$

It should be noted, however, that Underhill (1980) and Sahade (1980) have suggested that Wolf–Rayet stars have near normal compositions on the basis of their position in the luminosity/temperature diagram. The apparent lack of hydrogen they claim to be due to a misinterpretation of the line profiles and to insufficient understanding of the photospheric spectrum.

2.7. Stellar Winds

Absorption lines which may be attributed to the Wolf–Rayet star occur as short-wavelength satellites of the emission lines, i.e. as part of P Cygni type line profiles. Such profiles are indicative of mass outflow, and the increase in radial velocity with decreasing excitation potential of the upper level of the line (see figure 2.12) is generally interpreted as an outwardly accelerated motion combined with an outwardly decreasing temperature. The C III 190.9 nm absorption feature has been used by Noerdlinger (1979) to obtain a mass loss rate for HD 68273 of 1.1 to $1.4 \times 10^{-4} M_\odot$ a^{-1}. Willis and Wilson (1979) similarly obtain a mass loss rate of $9 \times 10^{-5} M_\odot$ a^{-1} for the same star. Radio fluxes from 5.0 to 14.7 GHz give 3.7 to $4.0 \times 10^{-5} M_\odot$ a^{-1} from HD 68273 (Morton and Wright 1978, Seaquist 1976, Wright and Barlow 1975) while 5 GHz observations of HD 192163 suggest $3 \times 10^{-5} M_\odot$ a^{-1} (Dickel *et al* 1980). The average mass loss rate from the infrared excesses is $1.6 \times 10^{-5} M_\odot$ a^{-1} (Hackwell *et al* 1974). Conti (1978) quotes $2 \times 10^{-4} M_\odot$ a^{-1} as the mass loss rate for HD 192163 from ultraviolet fluxes. These are enormous mass loss rates, and one need only consider that HD 68273 has remained apparently unchanged for over a century, and must therefore have lost some 1% of a solar mass since first observed, to realise that even with a mass of $10 M_\odot$ such a mass loss cannot be sustained for long.

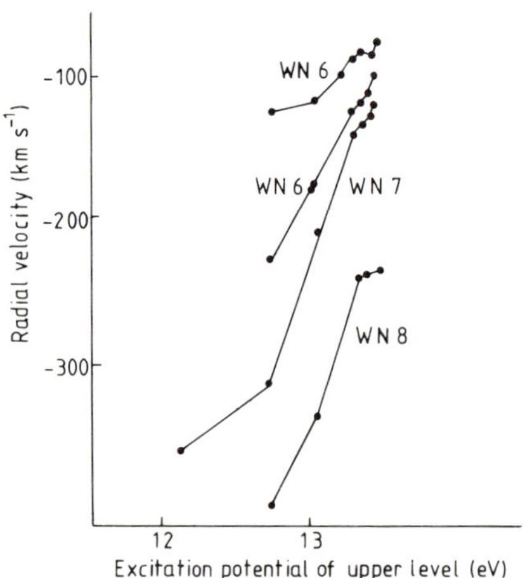

Figure 2.12. Mean displacements of the Balmer absorption line centres versus the excitation potential of the upper level of the line. (Reproduced from Niemela 1979 by permission.)

The initial mechanism for accelerating the wind is not clear; deposition of acoustic energy leading to a coronal type expansion, co-rotation due to magnetic fields or turbulence are all mechanisms that have been suggested (see chapter 1). A process based on stochastic instabilities in the velocity field has been suggested for Of stars by Andriesse (1980), and this could also be applied to Wolf–Rayet stars (see later in this chapter). The ultraviolet spectra reveal a correlation of terminal velocity with class for the WN stars – from 1300 km s^{-1} at WN8 to 3200 km s^{-1} at WN4 (Nussbaumer *et al* 1979). These observed terminal velocities are about three times the escape velocities from the stars (Abbott 1978), and may easily be achieved by a radiatively driven wind once the initial acceleration is accomplished. However, in V444 Cyg there seems to be a wind which expands at nearly constant radial velocity at large continuum optical depths (Hartmann 1978a). Such a velocity structure does not seem possible with radiatively driven winds and a different (and unknown) mechanism is required.

An interesting relationship exists between the number density of red supergiants (N_R) and the number density of Wolf–Rayet stars (N_{WR}) and their Galactocentric distance (see table 2.3). The effect may be attributable to the influence of heavy metal abundance on mass loss rates (Maeder *et al* 1980) and the length of time spent in these two stages.

Table 2.3. Ratio of number densities of red supergiants and Wolf–Rayet stars (Maeder *et al* 1980).

N_R/N_{WR}	Galactocentric distance (kpc)
0.14	7–9
1.5	9–11
13.0	11–13

2.8. Related Systems

2.8.1. Of Stars

The Of stars are closely related to the O stars, in that their absorption lines match those of stars from O3 to O9.5. However, they also have strong emission lines of N III 463.4, 464.0, 464.2 and He II 468.6. Subsidiary classes, O(f), and O((f)) have also been defined by Walborn (1971). O(f) stars have N III still in emission, but He II 468.6 is neutral or in weak absorption, while in O((f)) stars, N III is in emission, but He II is a strong absorption line.

Temperatures of Of stars lie between 30 000 and 55 000 K (Conti 1976), and their bolometric magnitudes between -9.5^m and -12^m. Several have been detected in the near infrared, and this emission (like that of the early Wolf–Rayet stars) seems likely to be due to free–free emission (Barlow and Cohen 1977).

Most Of binaries have mass ratios near unity, with both stars being large (20 to 30 $M_\odot$) (Conti *et al* 1980, Massey and Conti 1977, Morrison and Conti 1978), although it may be as low as 0.4 in UW CMa (Eaton 1978). The mass loss rates are generally found to lie near 10^{-5} to $10^{-6} M_\odot\ a^{-1}$, with all methods of measuring the mass loss rate in good agreement. For example, the 14.7 GHz observations of ζ Pup give $6.3 \times 10^{-6} M_\odot\ a^{-1}$, while the ultraviolet lines give $7.2 \times 10^{-6} M_\odot\ a^{-1}$ (Abbot *et al* 1980, Lamers and Morton 1976, Morton and Wright 1978). Kunasz (1980) has produced line profiles for a series of stellar wind models. A mass loss rate of $5.5 \times 10^{-6} M_\odot\ a^{-1}$ closely models the observed line profiles of Of stars. (HD 108 (O7If) and HD 148937 (O6.5f), however, only seem to be losing mass at a rate of $2 \times 10^{-7} M_\odot\ a^{-1}$ (Hutchings and Rudloff 1980).)

The close similarity between O and Of stars suggests from the theory of a radiatively driven stellar wind that their mass loss rates should also be similar (since this depends on the number of ultraviolet resonance absorption lines and not on the emission lines – see chapter 1). The mass loss rates discussed above for Of stars are, however, 4 to 100 times larger than those of comparable O type stars (figure 2.13) (Andriesse 1980, Conti and Garmany 1980, Lamers *et al* 1980). The mass loss rates in Wolf–Rayet stars are even higher (see earlier in this chapter) and in both cases are higher than the maximum loss rates expected for radiatively driven winds given by equation (1.64). A possible explanation for this excess mass loss may be that perturbations in the velocity field are unstable, and lead to stochastic instabilities in the flow (Andriesse 1980). In effect a third pressure is added to the thermal and radiation pressures, which adds a component to the mass loss rate given by (Andriesse 1980),

$$\dot{M} \simeq L_*^{3/2} \left(\frac{r_*}{M_*}\right)^{9/4} G^{-9/4} \tag{2.25}$$

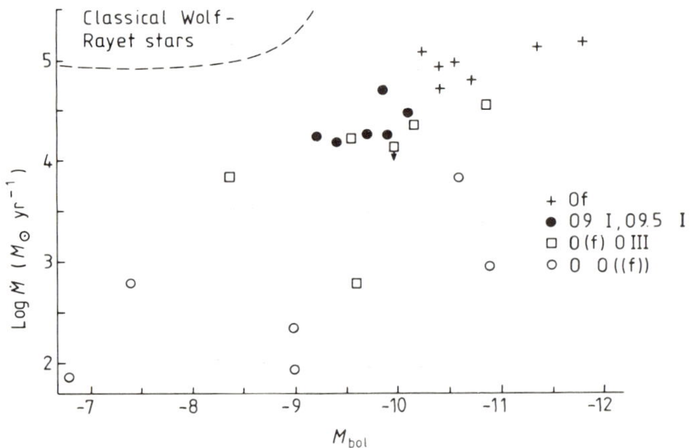

Figure 2.13. Mass loss rates in Wolf–Rayet, Of, and O type stars. (Reproduced from Conti and Garmany 1980 by permission.)

where r_* is the stellar radius and M_* is the stellar mass. Different mass/radius ratios will therefore lead to different mass loss rates for stars with the same temperature. There is a quasi-periodic 180 to 200 second flickering in the He II 468.6 line of HD 68273 which is easily interpreted in this model, and provides further support for it (Sanyal *et al* 1974).

An early Of star, HD 153919 (O5.5f), has been identified with the variable x-ray source, 2U 1700−37. Its optical and ultraviolet emission features vary in phase with the x-ray changes, whose period is 3.41 days (Hammerschlag-Hensberg 1978, Hammerschlag-Hensburg and Wu 1977, van Paradijs *et al* 1978, Walker 1974, Wolff and Morrison 1974). The C III/N III 464 nm feature is very strong, and Hβ has a P Cygni profile. The [Fe XIV] 530.3 line may be present in emission, but if so it is extremely variable (Lambert and Tomkin 1979). Other features, including the absorption components, are also phase dependent. An interpretation as a binary seems plausible, with the majority of the material lost as a stellar wind ($6.3 \times 10^{-6} M_\odot\, a^{-1}$), but with perhaps 4% of this accreting to the companion (Fahlman and Walker 1980). X-ray emission also occurs from ζ Pup (O4f) and may be explained by a 'lumpy' warm stellar wind (cf the stochastic mass loss process discussed above). In this, blobs of material are radiatively driven through the ambient gas and the resulting chaotic hypersonic motions produce temperatures up to 170 000 K and the observed x-ray emission (Hamann 1980, Lucy and White 1980). A similar picture is given by the ultraviolet and optical observations which imply changes in the mass loss rate by up to a factor of two in a few hours (Conti and Niemela 1976, Snow *et al* 1980). However, in another star, λ Cep (O6ef) the optical absorption lines and the ultraviolet P Cygni lines of C III and N V are stable, while the emission lines vary. Polarisation is also variable, with timescales of about a day. In this star an inhomogeneous flow near the star would therefore seem to be damped out and the flow stabilised at large radii (Brune *et al* 1979, Hayes 1978). ξ Per (O7.5IIIf) emits in the 0.15 to 4.5 keV region, with a peak near 1.0 keV, perhaps due to the same process as ζ Pup (Long and White 1980).

The Of and Wolf–Rayet stars lie close together just above the main sequence on the H–R diagram (figure 2.14). The N III emission lines are produced when the 3d levels become over-populated after dielectronic recombination from an auto-ionised N IV level (Conti 1976). They will occur in a plane-parallel atmosphere, as well as in an extended envelope. The He II line can only originate in emission from an extended envelope. Hence the Of stars seem likely to have large envelopes, while the O(f) and O((f)) stars may not necessarily do so.

A group of Of stars given a WN sub-class, and three WN stars given Of sub-classes by Leep (1979), shows the extent of the overlap between the groups. Conti (1976) said,

> 'Is there any essential difference between the Of stars and the Wolf–Rayet stars? I think that if one did not have any preconceptions . . . the spectra . . . would indicate only a difference of degree; the Wolf–Rayet star has a more extended envelope than the Of star.'

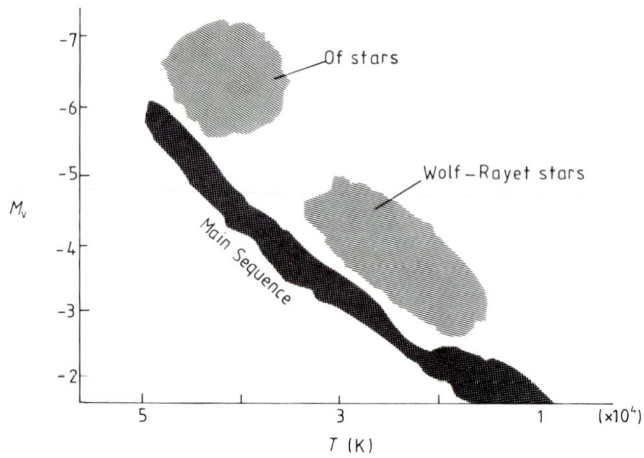

Figure 2.14. The upper region of the H–R diagram.

Both types of star are found physically associated. For example the un-named OB association at 17 h 15 min and $-38°$ contains two WN8 + OB systems, an O4f star and an O5f star, among its 15 members (Havlen and Moffat 1977). It seems reasonable then, to link the physical phenomena of Wolf–Rayet and Of stars, as the same processes operating to greater or lesser extents. There may also be an evolutionary link in the sense,

O star–O((f))–O(f)–Of–WN–WC star

with the envelope dimensions increasing and the hydrogen/helium ratio decreasing along the sequence (Andriesse 1980). Additional support for this view is given by the stochastic mass loss process proposed for Of stars, which can also explain the even higher mass loss rates of Wolf–Rayet stars if there are even further evolved O type stars. Stars which may be intermediate between Of and WN7 are HD 15570 and SK-65-22 in the Large Magellanic Cloud. Their spectra contain normal Of lines plus the N IV 148.6 and He II 164.0 emission lines which are more characteristic of Wolf-Rayet stars (Nandy *et al* 1980, Willis and Stickland 1980). Other links include HD 163758 (O7If) which appears to be carbon-rich (Leep 1978), and the Of binary BD + 40 4220 which seems to contain a Wolf–Rayet star in Of star clothing (Bohannan and Conti 1976).

2.8.2. P Cygni Stars

The link between Wolf–Rayet and P Cygni stars is much more tenuous than that with the Of stars, even though P Cygni type line profiles are found in most Wolf-Rayet spectra. Such line profiles are usually indicative of stellar winds and it is probably better to speak of a P Cygni phenomenon occurring in many different types of systems. The stars labelled P Cygni type are a diverse and interesting group which are discussed in detail in chapter 6.

2.9. Evolution

The data sketched above may, at least partially, be combined into a consistent picture by taking the view that Wolf-Rayet stars are the almost pure helium cores of stars originally having masses greater than 25 $M_\odot$. In the WN sequence the stellar composition is enhanced in nitrogen, by products of the CNO cycle appearing at the surface. In the WC sequence there is enhanced carbon due to products of the triple-alpha helium burning reactions coming to the surface. The original high mass is reduced by the stellar wind to that expected for Wolf-Rayet stars (which also removes the outer layers of the star revealing the helium core). The proportion of Wolf-Rayet stars to main sequence stars with masses greater than 25 $M_\odot$ is about one to four. With a Wolf-Rayet lifetime of about 10^6 years, and the main sequence lifetime of a 25 $M_\odot$ star of about 5×10^6 years, it appears likely that all stars with an original mass greater than 25 $M_\odot$ will become Wolf-Rayet stars (Smith 1973).

A substantial proportion, but not all, Wolf-Rayet stars are binaries. Hence a binary nature cannot be fundamental to their formation although it may aid mass loss (Vanbeveran and Doon 1980). (The apparently single Wolf-Rayet stars have a mean distance from the Galactic plane of 130 pc, while for the binary Wolf-Rayets it is 80 pc. The difference may arise from the single-line Wolf-Rayet stars originating as the secondary in a binary system whose primary undergoes a supernova explosion and expels the Wolf-Rayet star (Moffat and Isserstedt 1980).)

With this scenario, it appears that the WC stars may be more highly evolved than WN stars. This is supported by the presence of hydrogen in the late WN stars and their apparent link with the Of stars. The number densities of red supergiants are linked with those of Wolf-Rayet stars and their Galactocentric distance as indicated earlier (table 2.3). Further, the sum of their densities is a constant fraction (about 0.15) of the blue supergiant number density (Maeder *et al* 1980). We may perhaps therefore see the red supergiants as Of star precursors instead of the O type stars suggested earlier.

Post Wolf-Rayet phases are the subject of much speculation. Supernova precursors are a possibility (Moffat and Seggewis 1979), and links with pulsars or x-ray binaries have also been suggested (van den Heuvel 1976).

2.10. Planetary Nebulae Nuclei

The Wolf-Rayet stars which are the nuclei of planetary nebulae are obviously very different from the other two classes, despite similar spectra (see also chapter 3). In a similar manner to P Cygni stars it may be better to speak of a Wolf-Rayet phenomenon in these cases, with a very hot supersonic, outwardly accelerating and cooling stellar wind producing the spectrum. The nuclei of planetary nebulae are classified into:

Wolf-Rayet type,
Of type,
O type,
Continuum type,

so that not all planetary nebulae nuclei are Wolf–Rayet stars, however the link with the Of stars appears again (Gurzadyan 1970). A detailed comparison of two WC9 stars, one a population I type Wolf–Rayet star and the other a planetary nebula nucleus has been undertaken by Smith and Aller (1971). Although the masses differed by a factor of about 10, and the luminosities by a factor between 3 and 16, the line widths differed by only a factor of about 1.6, and the line strengths by factors of 1.4 to 2.6, so that similar values are required for the temperatures and surface gravities of the two stars. A model similar to that for the other types of Wolf–Rayet star seems adequate to explain the planetary nebulae nuclei, provided that the sizes of the envelopes are scaled down to the smaller sizes of these stars. Radiation pressure again seems likely to be the accelerating mechanism.

A peculiar object which mimics a Wolf–Rayet star is HD 45166. This seems to be a binary with a hot, one solar mass star which is losing mass at $5 \times 10^{-8} M_{\odot}\ \mathrm{a}^{-1}$, and a late B type star. Its small size leads to densities of ionised helium ($10^{17}\ \mathrm{m}^{-3}$) comparable with those in more normal Wolf–Rayet stars, despite its very low rate of mass loss (van Blerkom 1978a). The relationship (if any) of this star to other types of Wolf–Rayet star is not clear.

Table 2.4. Wolf–Rayet stars of eleventh magnitude or brighter (Smith 1968a, Westerlund and Smith 1964).

Star	Right Ascension$_{2000}$ h min	Declination$_{2000}$ ° ′	Magnitude (v)	Spectral type
HD 4004	00 43.4	+ 64 47	10.5	WN5
HD 9974	01 39.0	+ 58 09	10.8	WN3
HD 16523	02 41.2	+ 56 44	10.6	WC6
HD 32228	04 56.6	− 66 29	9.8	WC6 + O8
HD 269546	05 26.8	− 68 50	9.9	W + B3Ip
HD 50896	06 56.1	− 23 40	6.9	WN5
HD 62910	07 45.0	− 31 55	10.6	WN6 − C7
HD 63099	07 45.8	− 34 20	11.0	WC6 + O7
HD 68273	08 09.6	− 47 21	1.7	WC8 + O7
HD 76536	08 55.0	− 47 36	9.4	WC6
HD 86161	09 54.9	− 57 43	8.4	WN8
HD 90657	10 26.6	− 58 39	9.8	WN4 − C + OB
HD 92740	10 41.2	− 59 40	6.4	WN7
HD 92809	10 41.7	− 56 46	9.7	WC6
HD 93131	10 44.0	− 60 07	6.5	WN7
HD 93162	10 44.2	− 59 43	8.2	WN7 + O7
HD 94546	10 53.7	− 59 31	10.7	WN4 + BOn
MR 32†	10 58.7	− 61 11	10.9	WN8 + OB
HD 96548	11 06.3	− 65 30	7.9	WN8
HD 97152	11 10.0	− 60 59	8.3	WC7 + BOV
HD 97950	11 15.1	− 61 16	8.8	OB + WN
HD 104994	12 05.4	− 62 02	11.0	WN3
HD 113904	13 08.2	− 65 18	5.7	WC6 + O9.5
HD 115473	13 18.5	− 58 09	10.0	WC5
HD 117688	13 33.5	− 62 19	10.9	WN6 − C
HD 119078	13 43.3	− 67 24	10.1	WC7

Table 2.4. – *Continued.*

Star	Right Ascension$_{2000}$ h	min	Declination$_{2000}$ °	′	Magnitude (v)	Spectral type
HD 136488	15	24.2	−62	40	9.4	WC9
HD 137603	15	29.8	−58	35	10.2	WC9 + OB
HD 143414	16	03.8	−62	41	10.2	WN6
HD 151932	15	52.3	−41	51	6.6	WN7
HD 152270	16	54.3	−41	50	7.0	WC7 + O5 − 8
HD 156327	17	18.4	−34	24	9.7	WC7 + BOV
HD 156385	17	19.5	−45	38	7.5	WC7
HD 157451	17	25.4	−43	30	10.6	WC9
HD 164270	18	01.8	−32	43	9.0	WC9
LS 14‡	18	05.5	−23	09	10.2	OB + WN
HD 165688	18	08.0	−19	24	10.2	WN6
HD 165763	18	08.5	−21	15	8.3	WC5
HD 168206	18	19.1	−11	38	9.4	WC8 + BO
HD 186943	19	46.3	+28	16	10.4	WN4 + B
HD 187282	19	48.6	+18	12	10.6	WN4
HD 190918	20	06.0	+35	48	7.5	WN4.5
HD 191765	20	10.3	+36	11	8.3	WN6
HD 192103	20	11.9	+36	12	8.5	WC8 + OB
HD 192163	20	12.1	+38	21	7.7	WN6
HD 192641	20	14.6	+36	39	8.2	WC7 + Be
HD 193077	20	17.0	+37	26	8.2	WN5 + OB
HD 228766	20	17.5	+37	19	9.3	WN7 + O
HD 193576	20	19.5	+38	44	8.3	WN5 + O6
HD 193793	20	20.5	+43	51	7.2	WC7 + O5
HD 193928	20	21.6	+36	55	10.2	WN6 + OB
HD 197406	20	41.3	+52	35	10.5	WN7
HD 211853	22	18.7	+55	07	9.2	WN6 + BO
HD 214419	22	36.8	+56	54	8.9	WN7 + O7
HD 219460	23	15.1	+60	28	10.0	WN4.5 + BO

† MR – Number assigned by M Roberts (Bailey 1978).
‡ LS – Number assigned by L Smith (Abbott *et al* 1980).

3. Planetary Nebulae

3.1. Morphology

Very few planetary nebulae deserve the name in the sense that they appear planet-like when viewed through a small telescope. At least half appear star-like, and are only discovered to be planetary nebulae when their spectrum is examined. The morphology of some of those which may be resolved is shown schematically in figure 3.1. The shapes range from sharply bounded uniform discs, through rings, ellipses, bipolar and double, to beaded and irregular shapes. Gravitational braking, differential stellar rotation and radiation pressure can interact to reproduce most of these forms (Phillips and Reay 1977). In most nebulae a faint, intensely blue star is found near the centre. Their angular sizes are mostly up to 20″ to 40″ (although one, the Helix Nebula, NGC 7293, is $12' \times 15'$ and several of the large faint nebulae listed by Abell (1955) are over half a degree across). Their surface brightnesses are low; NGC 6720 (the 'Ring Nebula in Lyra') has a mean brightness of about 16^{m} per square second of arc, and this is one of the brightest examples.

Planetary nebulae which are non-circular and which have a principal axis show a correlation between this axis (or the one at right-angles to it) and the direction of their local interstellar magnetic field (Grimin and Zvereva 1968, Melnick and Harwit 1975). The correlation is strongest for those nebulae with the lowest Galactic latitude. The interstellar magnetic field is not strong enough to cause this alignment by itself (Heligman 1980) and some mechanism to amplify the magnetic effect may have to be invoked to explain the correlation. Alternatively, the observed correlation may be due to both the magnetic field and the cause of the asymmetry (rotation?) both being aligned to the structure of the galaxy.

High resolution radio observations of NGC 6543 and NGC 7662 at 5 GHz show that the radio structure is very similar to the optical appearance (Scott 1975). NGC 7027 however has a two-component radio source with the optical source coinciding with one of the components. The other radio component is observed in the infrared, but is presumably obscured in the visible (Terzian 1978).

About 1500 planetary nebulae have been identified to date, and useful lists of the brighter ones will be found in Acker and Marcout (1977), Lang (1978), Perek and Kohoutek (1967) and Weinberger (1977). They are concentrated towards the Galactic plane to about the same extent as Oort's intermediate population I (figure 3.2). That is, they are associated with evolved objects such as novae and RR Lyrae stars. The total number of planetary nebulae in the Galaxy is estimated to be about 4×10^5 (Osterbrock 1974).

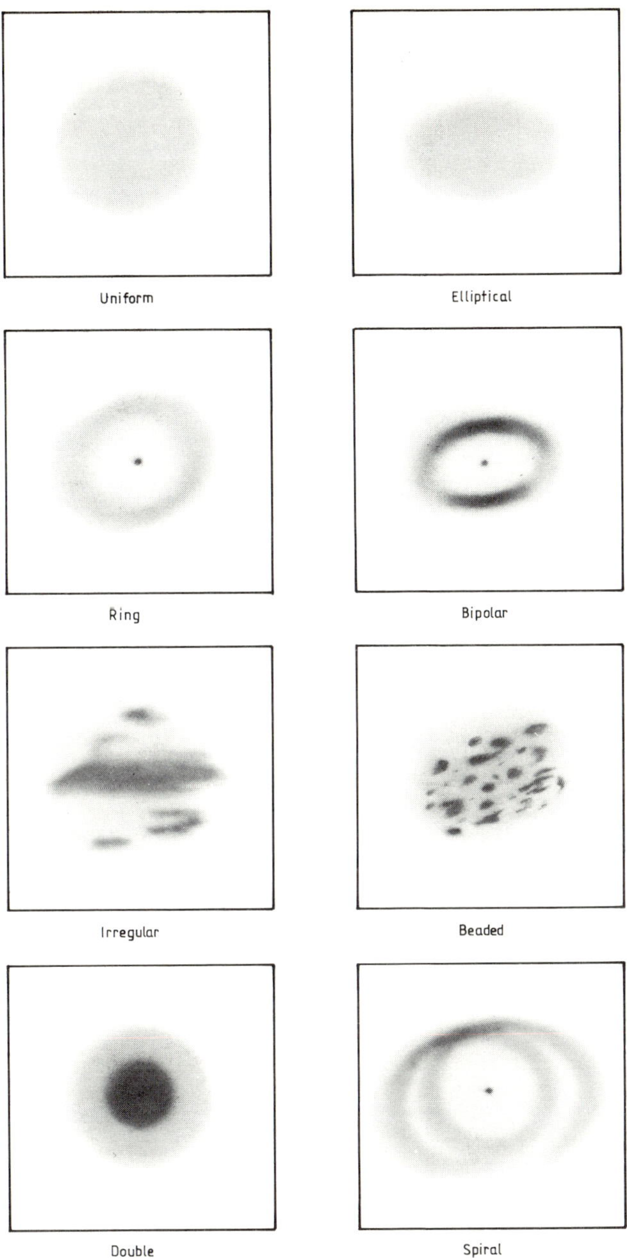

Figure 3.1. Some shapes of planetary nebulae.

3.2. Spectrum

The spectrum of the nebula contains three components (the spectrum of the central star is covered later in this chapter and also in chapter 2):

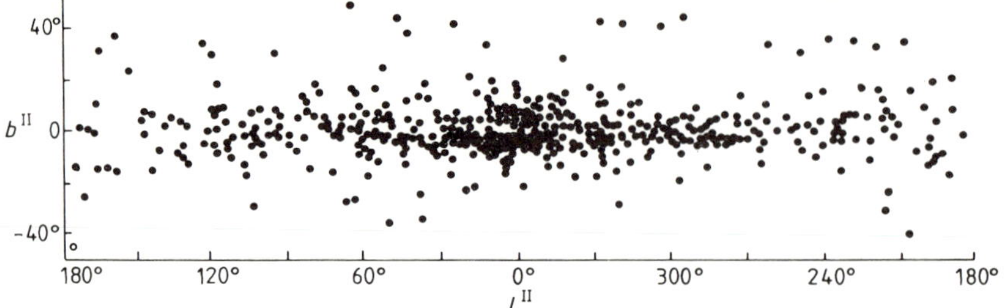

Figure 3.2. Distribution of planetary nebulae with respect to Galactic latitude and longitude. (From *Astrophysics of Gaseous Nebulae* by D E Osterbrock. W H Freeman and Co. Copyright © 1974.)

(a) forbidden emission lines,
(b) permitted emission lines,
(c) continuum.

This is shown schematically in figure 3.3, with the continuum in figure 3.4.

The nebulae can be classified on the basis of the excitation of their spectral lines. The line ratios that are used are (Aller 1969a).

$$A = \frac{[\mathrm{O\,III}]\ 372.7}{10 \times [\mathrm{O\,III}]\ 495.9} \tag{3.1}$$

$$B = \frac{[\mathrm{O\,III}]\ 372.7}{[\mathrm{O\,III}]\ 495.9} \tag{3.2}$$

$$C = \frac{[\mathrm{O\,III}]\ 500.7 + [\mathrm{O\,III}]\ 459.9}{10 \times \mathrm{H\,I}\ 486.1} \tag{3.3}$$

$$D = \frac{\mathrm{He\,II}\ 468.6}{\mathrm{H\,I}\ 486.1} \tag{3.4}$$

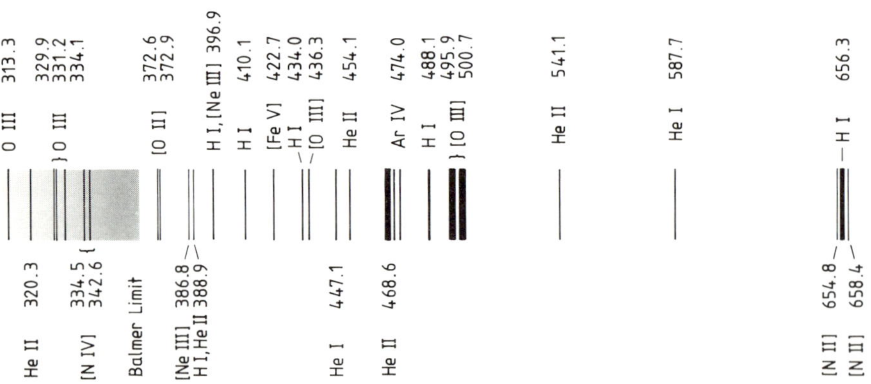

Figure 3.3. Schematic optical spectrum of a planetary nebula.

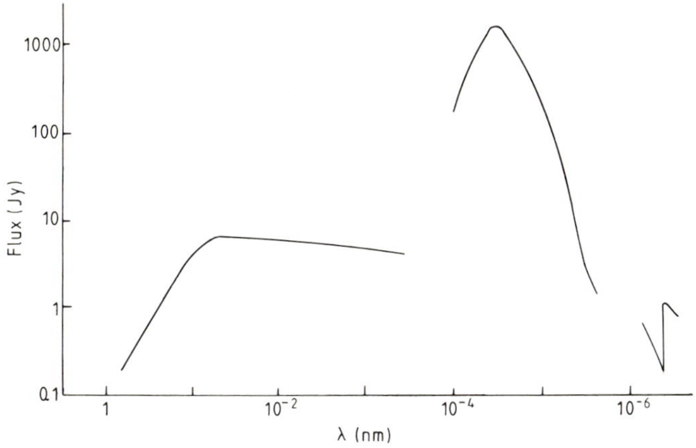

Figure 3.4. Continuous flux of NGC 7027. (Reproduced from Terzian 1978 by permission.)

$$E = \frac{[\text{Ne V}]\ 342.6 + [\text{Ne V}]\ 334.5}{[\text{Ne III}]\ 386.9} \tag{3.5}$$

and their variation with excitation class is shown in figure 3.5.

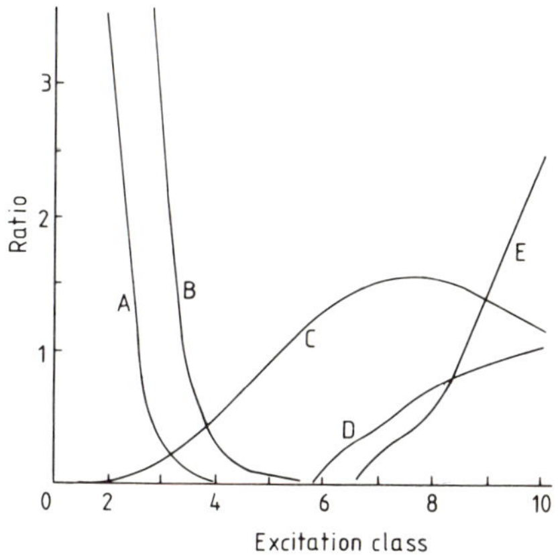

Figure 3.5. Planetary nebulae – excitation classes. (Reproduced from Aller 1969 courtesy of *Sky and Telescope*.)

The continuum radiation arises from a variety of sources; in the radio region it is due to free–free emission, in the infrared it is emission from dust particles heated by the central star, while in the blue part of the spectrum and in the ultraviolet it is due to recombination of hydrogen and helium (i.e. free–bound radiation). It was thought at one time that the ultraviolet continuum might be due to two-photon emission. This process allows a transition from the 2s level of hydrogen to the ground state (normally forbidden) by emitting two photons whose total energy is the same as that of Lyα. The transition probability is about 8 s^{-1}, and the process is not now generally thought to be significant.

Most of the permitted emission lines arise from the recombination of an ion and an electron, and the subsequent cascade back to the ground level. The majority are due to hydrogen, helium, carbon, nitrogen and oxygen. A few permitted lines of O III and N III rise through selective fluorescence with helium. The Balmer decrement calculated from recombination theory, including collisional effects, shows an excellent agreement with the observations (table 3.1). At one time there appeared to be a discrepancy between the observations and the theory for very high (n = 20 to 30) members of the Balmer series. But this turned out to be an error in photometric calibration of photographs of the spectra for faint lines.

Table 3.1. Balmer decrement in planetary nebulae (Brocklehurst 1971, Miller 1974).

	Hα	Hβ	Hγ	Hδ	H15	H16	H17	H18	H19
Observed (NGC 7027)	281	100	47.3	25.0	1.61	1.40	1.22	1.11	0.94
Calculated	279	100	47.2	26.2	1.68	1.41	1.20	1.03	0.90

The Balmer decrement would be much steeper if collisional processes played any part in producing the lines, so that recombination is undoubtedly the primary production process for the permitted lines. (In more distant nebulae, the Balmer decrement may be steeper than expected, but this is due to interstellar absorption.)

The forbidden lines mostly result from the population of metastable levels within a few electron volts of the ground state by collisions. Most of the lines are due to oxygen, neon and sulphur. They are so intense compared with the permitted lines because the electron energies are only a few electron volts. There is therefore insufficient energy involved in collisional excitation to push electrons to the high levels required for the production of the permitted lines (e.g. the production of Hα requires an initial excitation of over 12 eV). The metastable levels close to the ground state have lifetimes of 10^{-2} to 10^4 seconds for spontaneous emission. Collisional de-excitation is not significant only because of the very low density ($N_e \simeq 10^9$ m^{-3}) which gives a mean time between collisions of about 10^6 seconds.

The degree of excitation varies enormously within the nebula, so that lines of N I and Fe VII as well as molecular radio lines of CO and OH may be found in the same spectrum (Mutson *et al* 1975, Terzian 1978). The highest ionisation occurs nearest the central star, and the degree of ionisation declines towards the edge of the nebula. The nebula is therefore optically thick in the ultraviolet as the most energetic radiation is used up in producing the high levels of ionisation near the star.

Ultraviolet spectra of IC 418 show strong emission lines; C IV 154.9, C III] 190.8, C II] 232.6, [O II] 247.0 and Mg II 280.0 (figure 3.6) (Gurzadyan 1978, Harrington *et al* 1980), and in high excitation nebulae He II 164.0 (Koppen and Wehrse 1980a,b) also appears. The abundance of carbon which is found ($\log(N_C/N_H) = 8.85 \pm 0.21$) is not significantly different from the solar value. In IC 4997 by contrast the ultraviolet data lead to carbon and oxygen abundances which are an order of magnitude

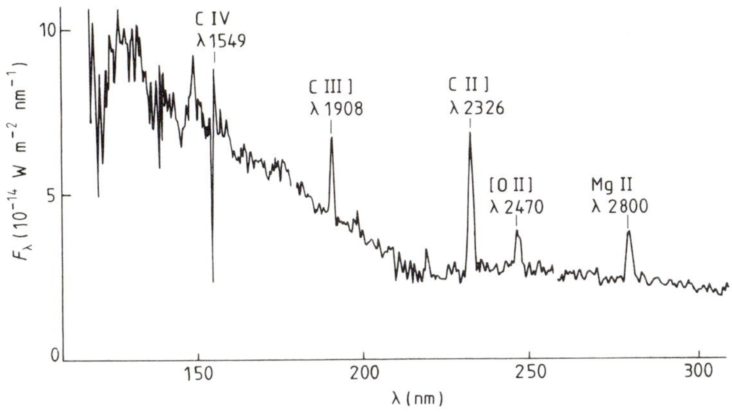

Figure 3.6. Ultraviolet spectrum of IC 418. (Reproduced from Harrington *et al* 1980 by permission.)

lower than in the Sun (Flower 1980). In NGC 7662 the [Ne IV] lines at 242.2 and 242.5 nm are found (Lutz and Seaton 1979) and lead to a value of the electron density of 1.1×10^{10} m^{-3} (see later in this chapter for a discussion of the methods of density determination).

The singly ionised forbidden lines of nitrogen and oxygen in high excitation nebulae are more intense than expected. This may reflect departures of the radiation field of the central star from that of a black body for wavelengths shorter than 91.2 nm, or it may indicate that charge exchange between neutral hydrogen and doubly ionised oxygen is important (Flower and Perinotto 1980). Alternatively a slight enrichment in helium and nitrogen has been suggested for some nebulae (Kaler 1979) which may contribute to the excess emission. Other abundance anomalies include the nebula 108-76.1° which is an halo object. Its helium abundance is normal, but oxygen is low by a factor of 25. In this case the abundance probably reflects the age of the object and its metal deficiency as a population II star (Boeshaar and Bond 1977).

A broad emission feature occurs at 3.3 μm in NGC 7027 which is probably attributable to dust grains (Tokunaga and Young 1980), as are some of the other features at 6.2, 7.7, 8.6 and 11.3 μm (Grasdalen 1979, Rank 1978, Russell *et al* 1977, Willner *et al* 1979). Features at 4.49 and 5.60 μm may be due to [Mg IV], [Mg V] or [Ar VI] (Russell *et al* 1977). In several nebulae the molecular hydrogen line at 2.122 μm is found (Beckwith *et al* 1978). Far infrared spectroscopy of NGC 7027 has revealed forbidden lines due to fine structure transitions. These are the lines of [Ne V] and [O IV] at 24.28 and 25.87 μm respectively (Forrest *et al* 1980). At even longer wavelengths, there seem to be two types of continuum for the nebulae. Most of the continua peak at about 35 μm and then decline more rapidly than a black body curve. A model which has optically thin dust at a temperature of about 100 K is suggested to explain this behaviour (McCarthy *et al* 1978). The other type of continuum is found in IC 418 and is much more sharply peaked.

The difference from other nebulae may arise from the grain composition and/or size, or be due to the geometry of the nebula (Moseley 1980). The composition of the grains may be carbon (Flower 1980) or iron-based (Scalo and Shields 1979, Shields 1975). (As mentioned earlier the carbon abundance from the ultraviolet lines in IC 4997 is very low, possibly due to the material condensing to form the dust grains.) It may be that the dust exists as two components: a cold dust cloud far out in the nebula which is responsible for the long-wavelength infrared emission, and a warmer component within the ionised region which emits the shorter wavelength radiation (Kwok 1980, MacGregor *et al* 1976).

A division of the planetary nebulae on the basis of their abundance has been made by Acker (1980), with a similar classification by Peimbert (1978). Those with the highest He/H and N/O enrichment are distributed in a very thin disc close to the plane of the galaxy, while those with the smaller ratios are akin to old disc–halo objects in their distribution and may be up to ten times older than the first group. A similar classification scheme due to Greig (1971, 1972), on the basis of the nebulae's appearance and spectra, recognises type B planetaries which have strong forbidden lines relative to Hα, and a filamentary appearance, while type C have forbidden lines which are weak compared with Hα, and increase in brightness towards their centres. (Type C planetaries are more or less equivalent to Acker's older grouping.)

3.3. Temperature

A close study of some of the levels involved in some of the forbidden transitions shows that some of the line ratios will be sensitive indicators of temperature (and also of density, as discussed in the next section of this chapter).

N II and O II have similar level diagrams (figure 3.7). For low densities ($N_e < 10^{11}$ m^{-3}) we may neglect collisional de-excitation, so that every excitation to the ^1S level of oxygen results in the emission of [O III] 436.3 or [O III] 232.1 (the same comments apply to N II with the appropriate substitutions of wavelengths). The relative probabilities of these two transitions will be given by their transition probabilities. Similarly every excitation to the ^1D level produces [O III] 495.9 or [O III] 500.7, again with their relative probabilities given by their transition probabilities. The transition producing the [O III] 436.3 lines also populates the ^1D level, but this is negligible in comparison with the direct process. Thus from Osterbrock (1974) we have the ratio of emission line strengths:

$$\frac{I_{495.9} + I_{500.7}}{I_{436.3}} = \frac{\Omega(^3P, {}^1D)(A_{^1S,{}^1D} + A_{^1S{}^1P})\,\bar{\nu}(^3P, {}^1D)}{\Omega(^3P, {}^1S)\,A_{^1S{}^1D}\,\nu(^1S, {}^1D)} \exp\left(\frac{\Delta E}{kT}\right) \tag{3.6}$$

where I is the intensity, A is the spontaneous transition probability, Ω is the collisional strength, ΔE is the energy difference between the $^1D_2$ and $^1S_0$ levels and $\bar{\nu}(^3P, {}^1D)$ is given by;

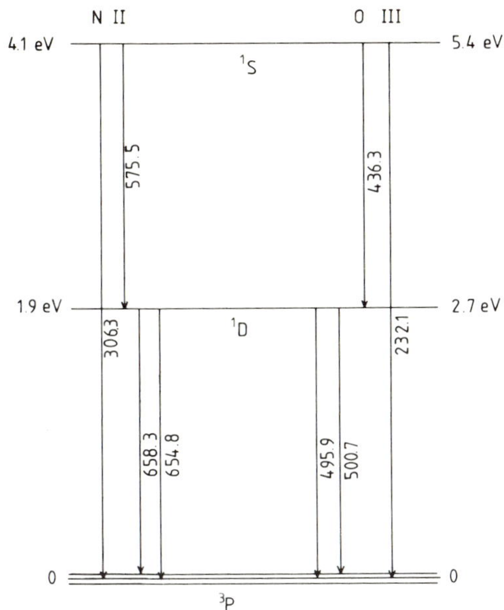

Figure 3.7. Schematic partial Grotrian diagrams for N II and O III.

$$\bar{\nu}(^3\mathrm{P}, ^1\mathrm{D}) = \frac{A_{^3\mathrm{P}_2, ^1\mathrm{D}_2}\, \nu_{500.7} + A_{^3\mathrm{P}_1, ^1\mathrm{D}_2}\, \nu_{495.9}}{A_{^3\mathrm{P}_2, ^1\mathrm{D}_2} + A_{^3\mathrm{P}_1, ^1\mathrm{D}_2}}. \tag{3.7}$$

This may be corrected to a first-order approximation by multiplying the right-hand side of equation (3.6) by a correction factor, f, where

$$f = \frac{1 + C(^3\mathrm{P}, ^1\mathrm{D})\, C(^1\mathrm{D}, ^1\mathrm{S})/C(^3\mathrm{P}, ^1\mathrm{S})\, A_{^1\mathrm{D}, ^3\mathrm{P}} + C(^1\mathrm{D}, ^3\mathrm{P})/A_{^1\mathrm{D}\, ^3\mathrm{P}}}{1 + [C(^3\mathrm{P}, ^1\mathrm{S}) + C(^1\mathrm{S}, ^1\mathrm{D})]/(A_{^1\mathrm{S}\, ^3\mathrm{P}} + A_{^1\mathrm{D}\, ^3\mathrm{P}})}, \tag{3.8}$$

where

$$C(i, j) = 8.63 \times 10^{-6}\, \frac{N_\mathrm{e}}{T^{1/2}}\, \frac{\Omega(i, j)}{\omega_i} \tag{3.9}$$

and ω_i is the statistical weight of level i. This gives

$$\frac{I_{495.9} + I_{500.7}}{I_{436.3}} = \frac{8.32 \exp(3.92 \times 10^4/T)}{1 + 4.5 \times 10^{-10}\, N_\mathrm{e}/T^{1/2}} \tag{3.10}$$

for the variation of line intensities with temperature. This is plotted for the low-density limit (i.e. N_e negligible) in figure 3.8. A similar equation may be found for N II;

$$\frac{I_{654.8} + I_{658.3}}{I_{575.5}} = \frac{7.53 \exp(2.5 \times 10^4/T)}{1 + 2.7 \times 10^{-9}\, N_\mathrm{e}/T^{1/2}} \tag{3.11}$$

which is also plotted in figure 3.8. Temperatures of a number of nebulae obtained by these methods are listed in table 3.2.

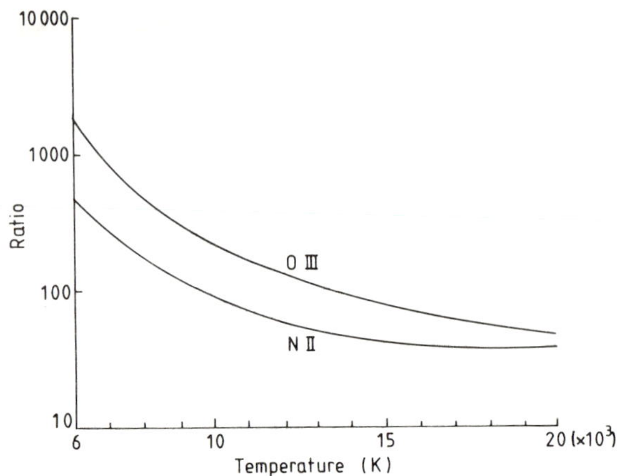

Figure 3.8. Variations in the line ratios: $(I_{495.9}+I_{500.7})/(I_{436.3})$ in O III and $(I_{654.8}+I_{658.3})/(I_{575.5})$ in N II.

Table 3.2 Temperatures of planetary nebulae (K) (Chaisson and Malkan 1976, Cohen *et al* 1977, George *et al* 1974, Gopal-Krishna 1978, Osterbrock 1974, Scott 1975, Weinberger 1977, Willis *et al* 1974).

Nebula	Infrared colour temperature	[N II] ratio	[O III] ratio	Radio continuum	Recombination line/continuum ratio
NGC 1535			12 000		
NGC 2392			18 400–18 800	11 000	
NGC 3242			12 400		11 300
NGC 6445			13 800	6 000	
NGC 6537			20 600	$<8 \pm 4 \times 10^4$	
NGC 6543		8 500	7 000–8 100	6 000–7 900	6 900–9 600
NGC 6572		12 900–15 500	10 600–11 000		6 900–11 500
NGC 6720			10 900	3 500	
NGC 6803		10 300	9 800		
NGC 6826			12 400		
NGC 7009			10 200–10 600	4 000–6 500	7 700–15 300
NGC 7027		11 500–14 300	15 300	14 000	20 400
NGC 7662			13 500–14 200	6 000	12 700–21 200
IC 418		8 200–10 000	10 000–11 500		7 700–17 900
IC 3568			11 400		
IC 4593			9 100		
IC 5217			11 700		
A 30	1000				
A 78	1000				

An infrared colour temperature has been obtained for two nebulae, A 30, and A 78, from their 2.28 to 3.5 μm emission (Cohen *et al* 1977). Since this will be the temperature of the dust, it is unlikely to be closely related to the temperatures obtained by other methods, and indeed, it results in temperatures of 1000 K compared with 6000 to 20 000 K (see table 3.2).

The temperature may also be obtained from the radio continuum radiation. The nebula will be optically thick for observations at long enough wavelength. Its emission at that wavelength (and longer wavelengths) will follow a black body curve, and so the brightness will be proportional to the square of the frequency, while at shorter wavelengths, where the nebula is optically thin, the brightness is almost constant with frequency. This latter, quite surprising, result arises from the almost inverse square dependence of absorption coefficient on frequency, while the energy density of radio waves by the Rayleigh–Jeans law (as already mentioned above) is:

$$\rho_\nu = \frac{8\pi k T \nu^2}{c^3} \qquad (h\nu \ll kT). \tag{3.12}$$

The product of optical depth and black body intensity is therefore nearly frequency independent. A typical intensity/frequency plot is shown in figure 3.9.

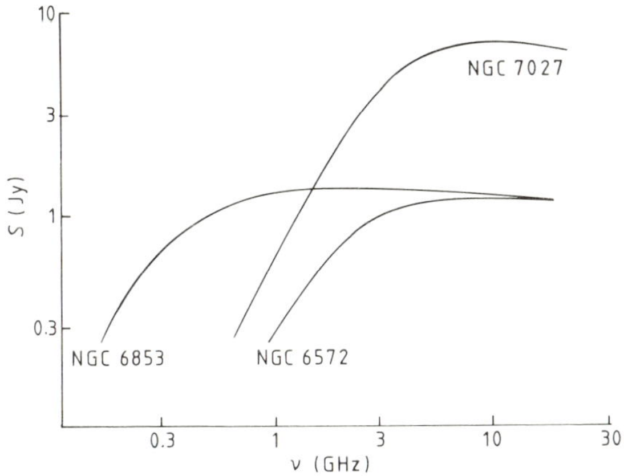

Figure 3.9. Radio spectra of planetary nebulae. (Reproduced from Terzian 1968 by permission.)

The surface brightness in the flat part of the spectrum is given by:

$$S_\nu \propto E T^{-1/2}, \tag{3.13}$$

where E is the emission measure, and is defined by

$$E = \int N_e^2 \, \mathrm{d}s \tag{3.14}$$

integrated across the nebula, while at long wavelengths it is given by:

$$S_\nu \propto T\nu^2. \tag{3.15}$$

Finally, the relative intensities of a recombination line and the recombination continuum may be used to determine temperature. The dependence on temperature arises from the dependence of the continuum emission intensity per unit frequency interval on the width of the electron velocity distribution curve. The Paschen continuum near Hβ and the Hβ line, or the Balmer continuum at the Balmer discontinuity and the Hβ line are usually used for this method. Their variations are plotted out in figure 3.10. Potentially the radio recombination lines may give a value for the temperature, but this method has primarily been applied to H II regions. Five planetary nebulae have had their H76α to H109α, and H113β, and He121α lines detected, leading to temperatures of 10 000 K (Terzian 1978, Terzian *et al* 1974). Radio recombination lines have also been observed in condensations in NGC 7027 (where they give a temperature of 19 000 K (Churchwell *et al* 1976)).

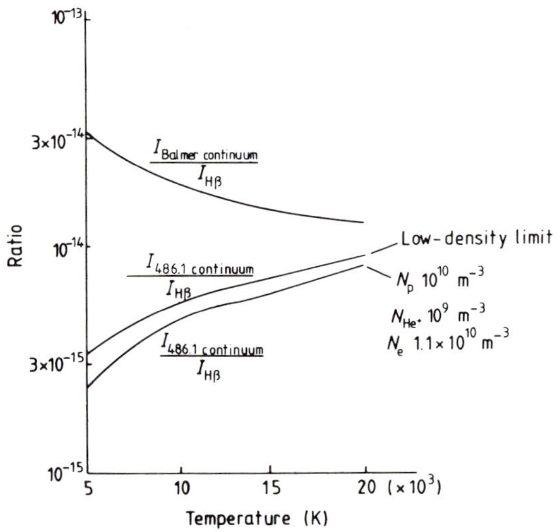

Figure 3.10. Ratio of continuum to line emission (Osterbrock 1974).

Temperatures of planetary nebulae obtained by these various methods are listed in table 3.2.

3.4. Densities

Density may be determined from line intensity ratios in a similar manner to temperature by a careful selection of lines. If the upper levels of two forbidden lines lie close to each other (in energy terms), then in the low-density limit, their line intensity ratio will be near unity. This is because the line strength will depend only on the level populations, and these will be in the ratio of their statistical weights, for sufficiently close lines populated by collisional excitation. At high densities (when the lifetimes of the levels are much longer than the mean time between collisions) the line strength will be proportional to the transition probabilities of the levels and their populations (again given by their statistical weights), so that the line intensity ratio will vary with density.

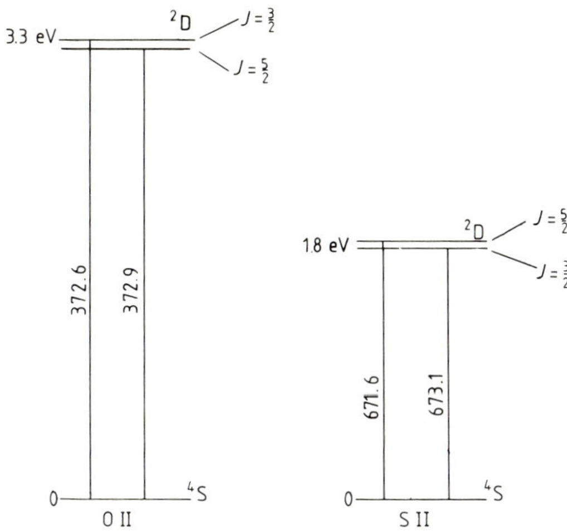

Figure 3.11. Partial Grotrian diagrams for O II and S II.

The lines usually employed are [O II] 372.9/[O III] 372.6 and [S II] 671.6/[S II] 673.1 (whose level diagrams are shown in figure 3.11). Other ratios used on occasion include; [O II] 732.5/[O II] 372.5, [S II] 407.0/[S II] 672.0, C III] 190.7/C III] 190.9 and [Cl III] 551.7/[Cl III] 553.7 (Aller 1969c). The ratios range from 1.5 in the low-density limit to 0.35 at high densities (see figure 3.12).

The density can also be estimated from the surface brightness provided that the distance of the nebula, its angular size and its filling factor (the proportion of the volume of the nebula occupied by radiating gas) are known. Hβ is usually measured for this purpose, and

$$I_{\mathrm{H}\beta} \propto N_{\mathrm{e}} N_{\mathrm{p}} R \tag{3.16}$$

where R is the radius of the nebula. If the nebula expands with constant mass then,

$$N_{\mathrm{e}} \propto R^{-3}, \tag{3.17}$$

so that,

$$I_{\mathrm{H}\beta} \propto N_{\mathrm{e}}^{5/3}. \tag{3.18}$$

Some typical densities are listed in table 3.3.

The electron density of NGC 7027 obtained from the H76α and H110α radio recombination lines is about $5 \times 10^{10}\ \mathrm{m}^{-3}$, which is in the range found from the line ratios (Chaisson and Malkan 1976).

The optical depth may be indicated by the [O I] 630.0/[O II] 372.7 and [N I] 519.9/[N II] 658.4 ratios. The former is 0.067 in optically thin nebulae and 0.019 in optically thick nebulae, while the latter ratio ranges from 0.013 to 0.003 in optically thin and thick nebulae, respectively (Kaler 1980).

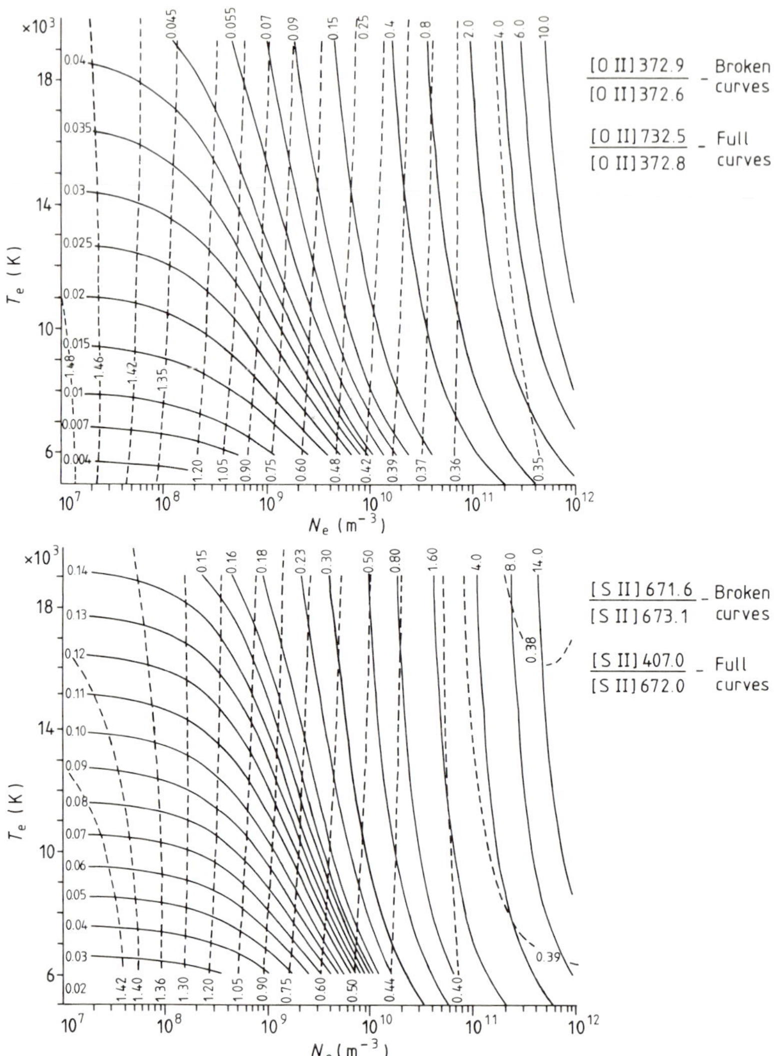

Figure 3.12. Variation in line intensity ratios with temperature and electron density. (Reproduced from Canto *et al* 1980 by permission.)

3.5. Distances

As a group, the mean distance of planetary nebulae is moderately well established. However the distances to individual nebulae are much less well known. Only one nebula – the Helix Nebula, NGC 7293 – is close enough for a trigonometrical distance to be found, and this is 20 to 30 pc (Osterbrock 1974). Several of their central stars have companions of recognisable spectral class, and so the spectroscopic parallax of this star gives a distance for the nebula.

A method which has been used on nearby nebulae, although it suffers from a number of possible sources of error, is based on the observation of radial expansion

Table 3.3. Densities of planetary nebulae (Fiebelman *et al* 1980, Kaler *et al* 1976, Lang 1978, Lutz 1974, Pottasch 1980).

	Electron density ($\times 10^9$ m^{-3})	
Nebula	Line ratio	Surface brightness
NGC 40	1.5	
NGC 1535	3.3	9
NGC 2346	0.4	
NGC 2371	4.2	
NGC 2392	3–9	6
NGC 2452	2	
NGC 3211	10	
NGC 3242	3–20	
NGC 6210	3.8–6	
NGC 6543	4.3–6.3	
NGC 6567	3	
NGC 6572	8–30	10
NGC 6720	0.8	1
NGC 6803	5.6	13
NGC 6818	3–10	
NGC 6826	1.6	6
NGC 6884	6.9	
NGC 6886	5.6	
NGC 6891	2.5	
NGC 7009	3.5–7	4
NGC 7026	9	
NGC 7027	10–400	
NGC 7662	4	
IC 418	30	20
IC 1297	30	
IC 1747	3.3	
IC 3568	6	0.8
IC 4593	2	2
IC 5217	8.5–20	6
BD + 30	7.2	
M 1–74	5.9	

of the nebula along, and perpendicular to, the line of sight. Most spectra of planetary nebulae are obtained by slit-less spectrographs and the spectrum consists of multiple images of the nebula, one for each of its emission lines. For the brighter nebulae however, high dispersion slit spectrograms may be obtained which, if the slit is placed symmetrically across the nebula, produce lenticular line profiles (figure 3.13). The maximum separation between the two components of such a line profile gives twice the expansion velocity of the nebula. Figures 3.14 and 3.15 show such

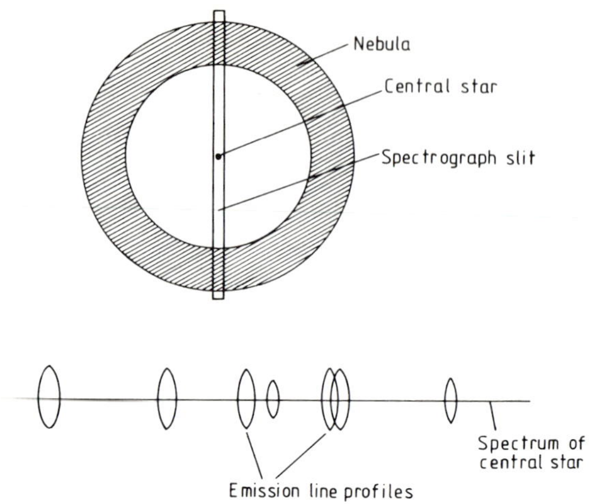

Figure 3.13. Schematic slit spectrogram of a planetary nebula.

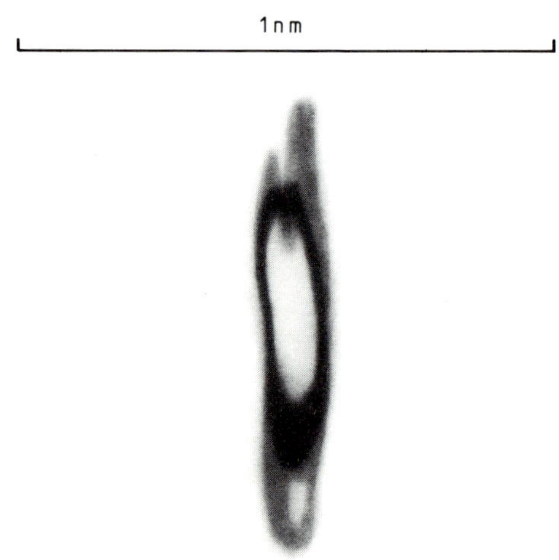

Figure 3.14. [O III] 500.7 line of NGC 7662 on a high dispersion spectrogram. (From *Astrophysics of Gaseous Nebulae* by D E Osterbrock. W H Freeman and Co. Copyright © 1974.)

a line and a profile across its centre. Direct photographs separated by a number of years allow the angular expansion to be determined from the change in angular size (figure 3.16). If δ is the observed rate of expansion in radius (in seconds of arc per year), and v the line of sight velocity of expansion (in km s^{-1}), then the distance, D, is given by,

$$D = 0.2v/\delta \qquad \text{pc.} \tag{3.19}$$

The expansion velocities range from about 0 to 60 km s^{-1}, with the mean at about 20 km s^{-1}.

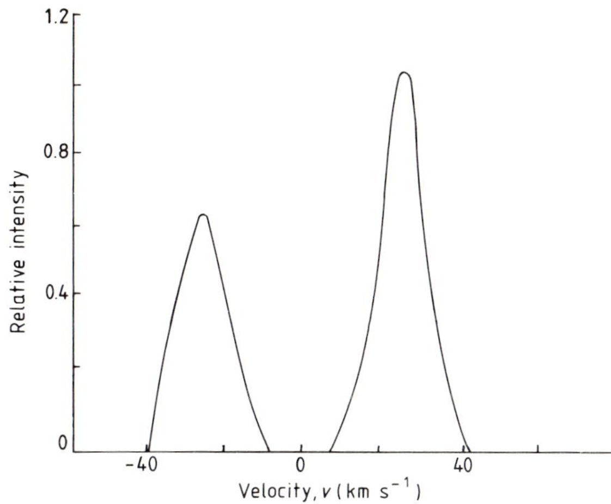

Figure 3.15. Profile of [O III] 500.7 in NGC 7662. (From *Astrophysics of Gaseous Nebulae* by D E Osterbrock. W H Freeman and Co. Copyright © 1974.)

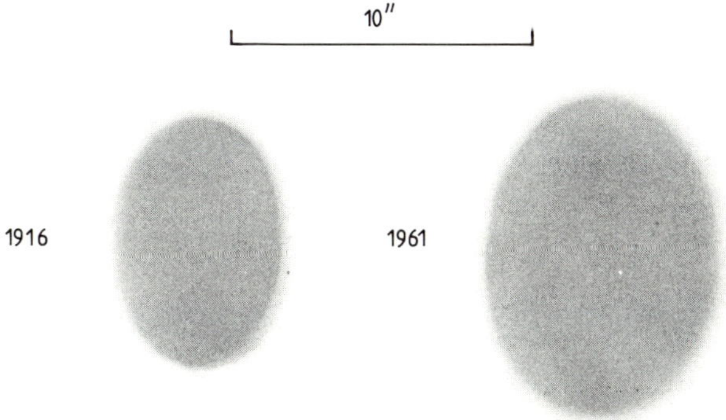

Figure 3.16. Change in the angular size of NGC 6572 between 1916 and 1961.

There are two main sources of error in this method. One arises from the assumption that the maximum observed line of sight velocity is equal to that perpendicular to the line of sight. As may be seen from the line profiles in figures 3.14 and 3.15, the velocities differ noticeably from uniform spherical expansion. The other error is that the observed angular expansion of the nebula may not be entirely due to physical expansion, but may also be due partly or wholly to a moving ionisation front.

Interstellar features in the spectrum of the central star may lead to potentially quite accurate determinations of the distances of planetary nebulae. This method can use the extinction dip at 220 nm, the ratio of radio free–free emission to the Hβ flux, or the comparison of theoretical and observed Balmer decrements. The method suffers from insufficiently well known interstellar extinctions and from difficulties in interpreting the spectrum of the central star. It therefore does not yet improve on the accuracy of other methods.

Two other indirect methods exist for obtaining the distances of individual nebulae. Firstly, one may find a nebula belonging to an object whose distance is known. M31 and the Magellanic Clouds provide the main examples of this method, although the nebulae are then too far away for many details to be made out. Furthermore, there seem to be systematic differences; the Large Magellanic Cloud planetaries have hotter central stars than those in the Small Magellanic Cloud, and both have oxygen abundances lower by a factor of two or three when compared with planetaries in the Galactic bulge (Webster 1976). The argon abundances of halo planetary nebulae are only about 1% of those of nearby planetaries (Barker 1980). (The abundances are generally close to the solar values (Barker 1980, Dinnerstein 1980) – see earlier in this chapter.) The comparison of distant nebulae with resolvable nebulae may therefore be unreliable. The second method is known as the Shklovsky indirect distance method (Osterbrock 1974). In this it is assumed that all nebulae have the same mass and are completely ionised. The radius is then inversely proportional to the cube root of the electron density (this method is the inverse of the surface brightness method for determining N_e). Since we have seen that

$$I_{H\beta} \propto N_e^{5/3}, \tag{3.20}$$

we have,

$$R \propto I_{H\beta}^{-1/5} \tag{3.21}$$

of, if θ is the angular radius of the nebula,

$$D \propto I_{H\beta}^{-1/5}\,\theta^{-1}, \tag{3.22}$$

and since

$$I_{H\beta} \propto F_{H\beta}/\theta^2 \tag{3.23}$$

where $F_{H\beta}$ is the observed flux in Hβ at the Earth, corrected for interstellar reddening, we have,

$$D \propto F_{H\beta}^{-1/5}\,\theta^{-3/5}. \tag{3.24}$$

The constant of proportionality will depend on the mass, composition and filling factor of the nebula, and is best determined by measurement of nebulae whose distance is known. With D in parsecs, $F_{H\beta}$ in W m^{-2} and θ in arc seconds, the constant is about 14, so that;

$$D = 14F_{H\beta}^{-1/5}\,\theta^{-3/5} \qquad \text{pc.} \tag{3.25}$$

This method generally seems to over-estimate the distance, probably because only the most extended nebulae are completely ionised (Pottasch 1980). The distances to planetary nebulae obtained by a variety of methods, are listed in table 3.4.

Table 3.4. Distances to planetary nebulae (Allen 1973, Liller 1978, Perek and Kohoutek 1967, Pottasch 1980).

Nebula	Distance (pc)	Nebula	Distance (pc)
NGC 40	620	NGC 6891	670
NGC 246	380–465	NGC 7008	790–840
NGC 1514	400–800	NGC 7009	620–700
NGC 1535	600	NGC 7026	1700–1800
NGC 2346	640–1250	NGC 7027	1200
NGC 2371	500	NGC 7293	20–30
NGC 2392	700–1000	NGC 7662	530–980
NGC 2452	2000	IC 418	1500
NGC 3132	670–800	IC 1747	1100–1200
NGC 3242	500–800	IC 3568	480
NGC 3587	470–600	BD + 30	800
NGC 3918	1200	M 1–59	1000–1430
NGC 6210	1500	He 2–131	700
NGC 6543	900	Un-named at 6 h 15 min 55.6°	140
NGC 6567	1000	Nebula associated with FG Sge	2500
NGC 6572	450–900		
NGC 6720	500–700		
NGC 6803	1000		
NGC 6826	800		
NGC 6853	220–320		

Proper motions are measured for the central stars of about 40 planetary nebulae, and these give statistical mean parallax. But since the nebulae move in elliptical orbits around the galaxy, they are high-velocity objects. Their velocity distribution is not random (figure 3.17) therefore, and such a measure is not very useful.

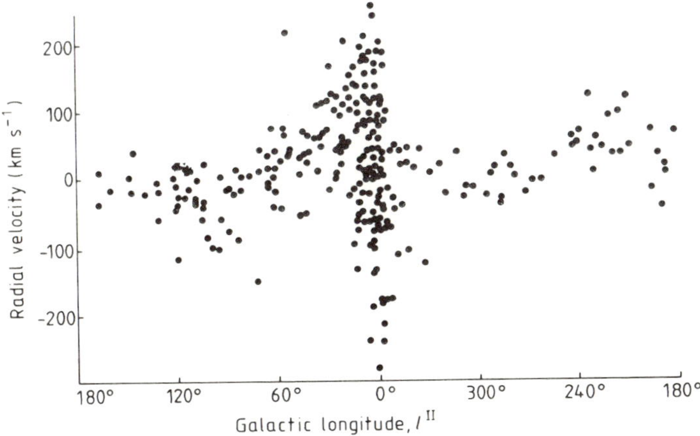

Figure 3.17. Radial velocity distribution of planetary nebulae. (From *Astrophysics of Gaseous Nebulae* by D E Osterbrock. W H Freeman and Co. Copyright © 1974.)

The results of the distance measurements show the nebulae to be concentrated towards the Galactic plane (figure 3.2). Their RMS height above the plane is about 215 pc, which suggests that they belong among the old population I type objects.

3.6. Masses, Dimensions and Luminosities

With the distance and the electron density known, it is simple to obtain the mass of ionised gas, size and luminosity. These are listed in table 3.5.

Table 3.5. Mass, diameter and absolute magnitude of planetary nebulae (Allen 1973, McCarthy *et al* 1978, Perinotto 1975, Pottasch 1980).

Nebula	Mass ($M_\odot$)	Diameter (pc)	Integrated absolute photographic magnitude
NGC 40	0.037–0.11		
NGC 246	0.12	0.4	+ 0.7
NGC 1514	0.095		
NGC 1535	0.0076		
NGC 2346	0.059		
NGC 2371	0.015		
NGC 2392	0.051–0.10	0.18	− 2.4
NGC 2452	0.098		
NGC 3132	0.06–0.12	0.20	− 1.3
NGC 3242	0.025–0.04	0.10	− 1.1
NGC 3587	0.10–0.14	0.5	+ 2.4
NGC 3918		0.08	− 2.0
NGC 6210	0.0014–0.13	0.08	− 1.5
NGC 6543	0.12	0.08	− 1.6
NGC 6567	0.0087		
NGC 6572	0.0084–0.10	0.05	− 2.0
NGC 6720	0.083–0.17	0.20	− 0.8
NGC 6803	0.012		
NGC 6826	0.08	0.10	− 0.8
NGC 6853	0.17–0.35	0.3	+ 0.9
NGC 6891	0.0055		
NGC 7008	0.2		
NGC 7009	0.03–0.09	0.08	− 1.1
NGC 7026	0.06–0.13		
NGC 7027	0.2–1.0	0.07	− 2.2
NGC 7293	0.19–0.34	0.5	+ 1.0
NGC 7662	0.049–0.07	0.06	− 1.7
IC 418	0.04–0.17	0.09	+ 0.2
IC 1747	0.016		
IC 3568	0.0012		
BD + 30	0.0068		
M 1–59	0.006		
He 2–131	0.0024		

The masses of nearby nebulae increase with size and with decreasing distance (Pottasch 1980). This appears to be due to the smaller nebulae being optically thick to Lyman continuum radiation. As the nebulae expand and their densities decrease a greater amount of the material is ionised, and so the observed ionised mass increases. The total mass of the nebula may be one or two orders of magnitude higher than the ionised mass except in the largest and most rarefied nebulae. The total mass is therefore likely to be 0.2 to 0.6 $M_\odot$ (Pottasch 1980). (The relationship between observed ionised mass and size may give another method for determining the distances of planetaries (Maciel and Pottasch 1980).)

There is a relationship between the integrated absolute magnitude of the nebula (M_n), and the magnitude of the central star, given by (Allen 1973)

$$M_n \simeq -1.5 + 0.8\delta \qquad (3.26)$$

where δ is the magnitude difference between the star and the nebula (i.e. $\delta = m_* - m_n$). It is related to the temperature of the central star as shown in figure 3.18.

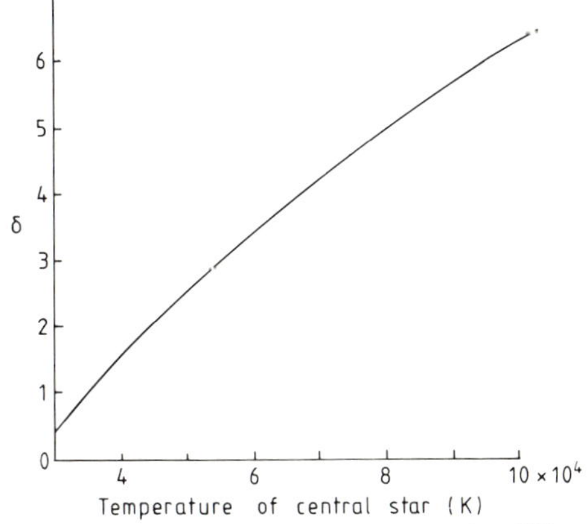

Figure 3.18. Relationship between δ (the magnitude difference between the central star and the nebula) and the temperature of the central star, for planetary nebulae.

The mean absolute magnitude of planetary nebulae in the Small Magellanic Cloud is about -5^m. This is appreciably brighter than those listed above, and reflects the observational bias towards bright nebulae because of the large distances to Small Magellanic Cloud nebulae.

As the nebula expands it becomes optically thin in additional regions of the spectrum allowing more of the radiation of the central star to escape, and so the nebulae become fainter with increasing radius (figure 3.19).

3.7. Fine Structure in the Nebulae

The fine structure may be considered under two headings; stratification and irregularities.

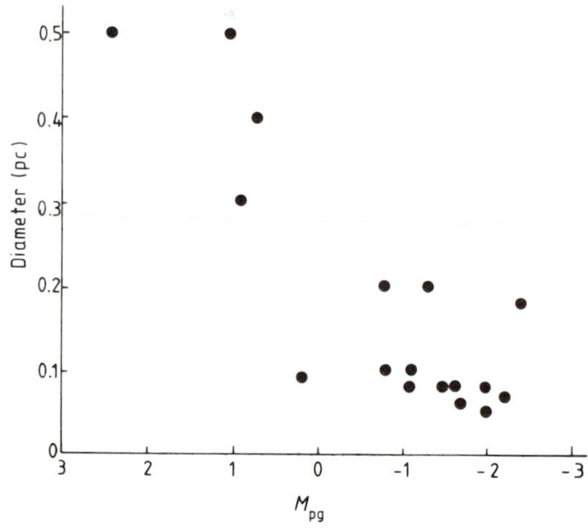

Figure 3.19. Variation of diameter with absolute magnitude in planetary nebulae.

3.7.1. Stratification

Slit-less spectrograms produce a series of monochromatic images, and these frequently show a variation in their diameter which depends largely upon the ionisation potential of the ion producing the line (table 3.6). The reason for the stratification lies in the number of quanta available to produce ions. The photon distribution for a black body at 50 000 K (a typical temperature for the central star of a planetary nebula) is shown in figure 3.20. The very high-energy quanta will produce high ions close to the star, but will rapidly be completely absorbed. Close to the star there will therefore be very few atoms in low stages of ionisation, so that the lower energy quanta will pass through largely unscathed, only ionising atoms further out

Table 3.6. Dimensions of monochromatic images of planetary nebulae (Gurzadyan 1970).

	Size (seconds of arc)				
Nebula	Hβ	[O III] 500.7+495.9	[Ne III] 386.8	He II 468.6	Ne IV 342.6
NGC 2165	5.3	5.7	5.9	4.4	3.4
NGC 2392	13.0	14.2	13.6	12.8	11.8
NGC 2440	5.7	6.4	6.5	5.3	3.1
NGC 6816	17.6	19.7	19.2	14.4	10.8
NGC 6886	3.9	5.0		3.4	3.1
NGC 7026	6.0	6.3	6.1	3.7	
NGC 7662	13.9	14.1	14.7	12.1	9.1
Ionisation potential (eV)	13.5	35.0	40.9	54.2	96.0

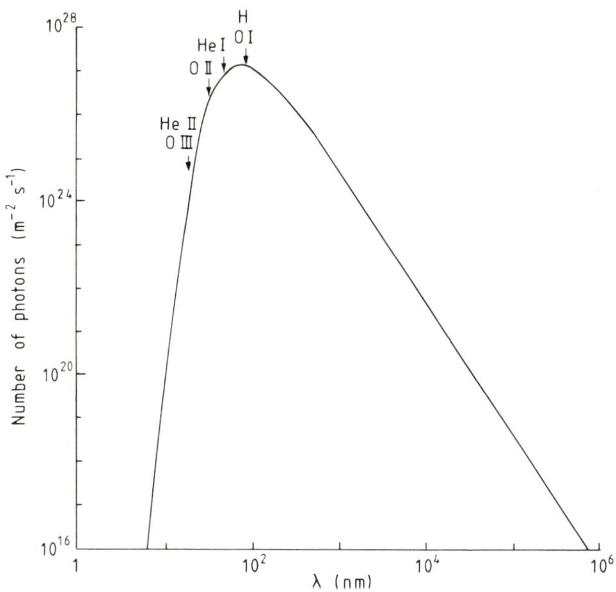

Figure 3.20. Number of photons per unit wavelength interval for a black body at 50 000 K. Wavelengths corresponding to the ionisation energies of hydrogen, helium and oxygen are indicated by arrows.

when all the high-energy quanta have been absorbed. Hydrogen and neutral oxygen have very similar ionisation levels, and so should occupy roughly the same region of the nebula. However since hydrogen can only be ionised once, it can recombine throughout the whole of its ionised region to produce emission lines. Hence the hydrogen shell will have the centre of its image more filled in with emission than the O II image (figure 3.21).

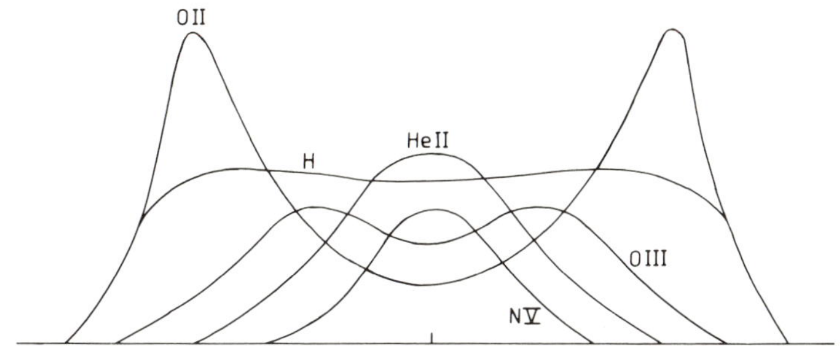

Figure 3.21. Schematic intensity profiles in the light of various ions from a planetary nebula showing stratification.

The ionised hydrogen region is known as a Stromgren sphere, and similar structures are found in many other types of nebula. The Stromgren radius is given by

(Stromgren 1939),

$$\log_{10} s = 3.56 - \frac{2.273 \times 10^4}{T} + \frac{1}{6} \log_{10}\left(\frac{T_n T R^2}{N^4}\right) \tag{3.27}$$

where s is the Stromgren radius (pc), T is the stellar temperature, T_n is the electron temperature in the nebula at s, R is the stellar radius ($R_\odot$) and N is the particle number density in the nebula. Similar relationships exist for other ions, but these will be modified by interactions with higher and lower ions of the same atom.

The sizes of the images in the forbidden lines should be larger than those in the permitted lines of the same ion. The permitted lines of an element are produced (generally) by recombination from a higher ion, and their images should therefore correspond to the region containing the higher ion, while the forbidden lines are produced by collisional excitation of the ion, and their images should correspond to the region containing the ion of interest. The images in permitted lines should therefore correspond to the images in the forbidden lines of the next stage of ionisation of the element.

As nebulae expand, the ionised region extends to larger and larger proportions of their mass so that the largest nebulae are completely ionised, and stratification is much reduced.

There are also nebulae in which there appears to be stratification into a spiral shape. The Helix Nebula (NGC 7293) is the best known example; another is NGC 6543. It is not clear whether or not this is really an helical structure. In the Helix Nebula the Hα data may be fitted to a fairly normally shaped shell, but the [N II] data require an helix with a pitch of 1/3, inclined at an angle of about 50° to the line of sight. In another model, the helix has the axial coordinate proportional to the square of the azimuthal angle (Taylor 1977). If the structure is truly helical, then this could arise through the gravitational focusing of the slowly moving gas by a distant companion. Precession of the orbit through mass loss induced torques will lead to helical, or at least curvilinear structures, for the nebula (Fabian and Hansen 1979). NGC 6853 (the 'Dumb-Bell Nebula') has large turbulent motions near the star, and may be explicable by a cylindrically expanding nebula (Goudis *et al* 1978).

3.7.2. Irregularities

Many planetary nebulae have small-scale irregularities in the form of bright or dark nodules within them or close by them. In some the disposition of these nodules has led to a recognisable structure (recognisable only with considerable imagination in some cases!). Thus we have NGC 7009 (the 'Saturn Nebula'), NGC 3587 (the 'Owl Nebula') and NGC 2392 (the 'Eskimo Nebula') – see figure 3.22. In others the nodules occur scattered throughout the nebula (see figure 3.1, 'Beaded type').

In NGC 7293 there are numerous fine radial filaments on the inner edge of the nebula. Their dimensions are typically 1 x 30 seconds of arc (corresponding to

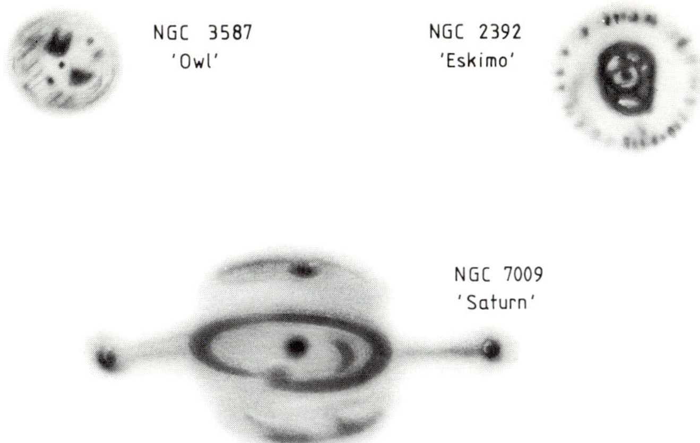

Figure 3.22. Planetary nebulae with recognisable structure.

2×10^{10} by 6×10^{11} km for a distance of 140 pc), and they are moving outwards at about 5 km s^{-1}. Intensified 'nebular' emission of neutral and singly ionised atoms occurs in the filaments, while 'auroral' lines are diminished. It is suggested that the filaments form from optically thick regions at the head of the filament (the end closest to the star) (van Blerkom and Arny 1972). Behind these regions the material is in a shadow and is only singly ionised, while the surrounding material is more highly ionised. The filaments will therefore be most visible in the light of singly ionised atoms. The optically thick regions may be Rayleigh–Taylor instabilities where the main mass of the nebula interacts with the radiation of the central star in a shock front.

3.8. Variability

The size of the nebulae (a light month to a few light years) precludes variations on short timescales. By direct photography only the slow expansion of the nebula can be seen. Spectroscopically, there has been a slow decrease in the ratio of [O III] 436.3 to Hα, from about 1.5 in 1895 to 0.6 in 1969, in the nebula IC 4997 (Aller 1969c). This is a small compact nebula of high surface brightness which may have formed as recently as 200 years ago. The relative effect of expansion at about 15 km s^{-1} on density and temperature etc is much larger on a small than on a large nebula. Similar changes in other nebulae will therefore probably have timescales of 10^3 years or more. Superimposed on this secular variation are shorter term (about one year) variations in the line ratio of up to 20% (Fiebelman *et al* 1979). Such variations may reflect changes in the emission of the central star. The ratio is sensitive to electron temperature, and an alteration by 5 to 10% in the number of ionising photons could produce the observed changes. M 2–9 has also shown variability, brightness changes being found with a timescale of about three years. Van den

Bergh (1974) has interpreted these variations as illumination of the nebula by shifting beams of ultraviolet radiation that escape between dust clouds orbiting the central star.

3.9. The Central Star

The central stars of planetary nebulae have already been mentioned in chapter 2 which discusses Wolf–Rayet stars. The spectra of the central stars were divided into four classes:

Wolf–Rayet type (including WR + Of)
Of type (including O VI stars)
O type
Continuum type.

There are also peculiar/unclassified stars which do not fit this scheme (Lutz 1978). Not all planetary nebulae have observable central stars, but there is no reason to expect that a central star does not exist in such cases. In the smallest and densest nebulae it may be obscured by the nebula, while in the largest nebulae it may be very faint and lost amongst the background and/or foreground stars. Some nebulae (e.g. He 2–36) have a central star with a spectral class between A and K (Lutz 1977, Mendez 1978). The excitation of the nebula however is much higher than would be expected from such a star. Almost certainly there is a hot companion which has yet to be discovered. NGC 1514, NGC 2346 and NGC 3132 once belonged in this class of nebulae, but since then the hot companions to their late stars have been observed (Kohoutek and Lausten 1977, Mendez *et al* 1978).

The spectra of the first two classes of central star have been described in chapter 2. (An extension of the WC sequence to lower temperatures exists for the central stars of M 4-18, He 2-113 and CPD $-56°$ 8032, which have strong C II lines and fainter He I, C III, O II, [O II], O III, Si III and Si IV lines (Webster and Glass 1974). The O type central star has a pure absorption line spectrum, with lines characteristic of spectra class O. The continuum type have featureless spectra with no emission or absorption lines. In some of the latter cases, the spectrum may not have been observed with adequate spectral resolution, so that absorption lines which are partly filled in by emission cannot be distinguished. But in other cases (e.g. NGC 3242) there is a genuine lack of spectral features.

The temperatures of the central stars have been determined by the Zanstra method. In this method, the nebula is assumed to be optically thick to ionising radiation, and in equilibrium. Every ionisation of hydrogen will then result in the emission of a Lyα photon, a Balmer photon, and possibly photons of higher series (Paschen, Brackett, Pfund, etc) as shown in figure 3.23. The Balmer photon, and higher series photons, escape from the nebula without further absorption in almost all cases. The Lyα photon slowly scatters its way out of the nebula. Thus the number of photons in the Balmer lines and in the Balmer continuum is equal to the

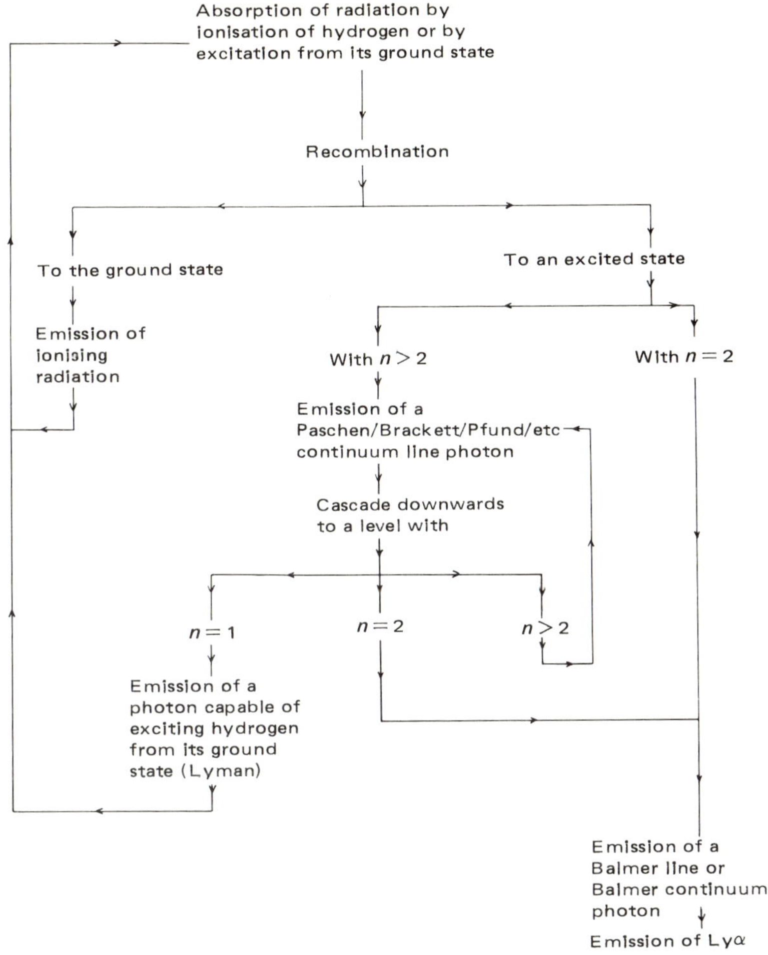

Figure 3.23. Ionisation/recombination cycle in hydrogen.

number of photons with wavelengths shorter than 91.2 nm. Since the Balmer decrement for a recombination spectrum is well known (table 3.1), it is only necessary to measure the number of photons in one Balmer line (usually Hβ) in the nebula, to determine the number of photons of wavelength less than 91.2 nm emitted by the star. A second measure of the number of photons in some part of the star's continuum will then give a long base-line colour index, and so a temperature.

For a black body distribution, we have;

$$\frac{N_{\nu_c}}{XN_{H\beta}} = \frac{\nu^2\,[\exp(h\nu_c/kT)-1]^{-1}}{\int_{3.3\times 10^{15}}^{\infty} \nu^2\,[\exp(h\nu/kT)-1]^{-1}\,d\nu} \tag{3.28}$$

where N_{ν_c} is the number of photons at some continuum frequency, ν_c, $N_{H\beta}$ is the number of Hβ photons, X is the ratio of total number of Balmer photons (line and continuum) to the number of Hβ photons for the conditions in the nebula, and T is the temperature of the central star. However the spectrum at short wavelengths is strongly affected by the ionisation edges of various elements (figure 3.24). The determination of the right-hand side of equation (3.28) has therefore to be undertaken from model atmosphere calculations, rather than by the assumption of black body emission.

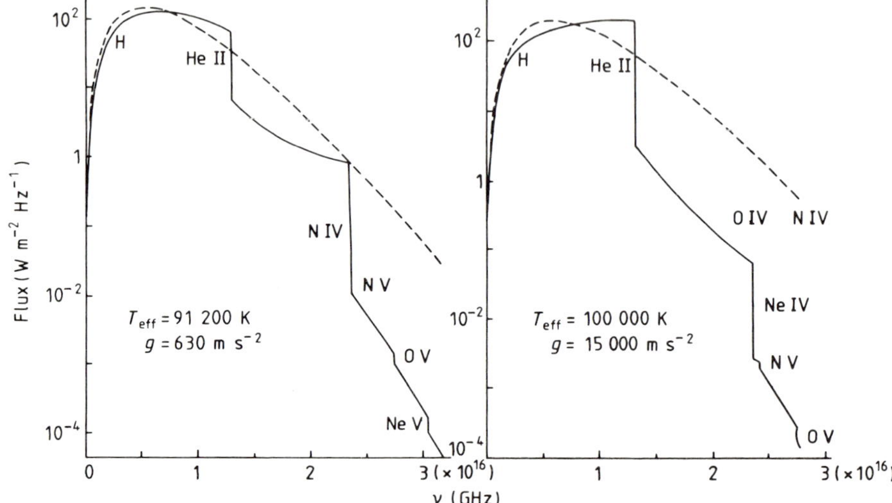

Figure 3.24. Comparison of model atmosphere (full curve) and black body energy distributions (broken curve) at high frequencies. The ionisation edges of various elements are indicated. (Reproduced from Bohm 1968 by permission.)

Since neutral oxygen and hydrogen have similar ionisation potentials, the presence of O I lines in the nebula suggests that neutral hydrogen will also be present. The assumption of optical thickness for radiation of shorter wavelengths than 91.2 nm is likely to be valid in such cases. If O I lines are not found, then a similar procedure can be carried out using lines of He II to determine the flux short of 22.8 nm, the ionisation limit of ionised helium. Similar procedures can be devised for other elements. The forbidden lines may also be compared with the recombination lines, to obtain a related temperature, since the electrons which collisionally excite the atoms and ions ultimately derive their energy from ionising photons.

Ultraviolet spectra of the central star of NGC 6543 show emission lines of C IV 155.0, He II 164.0 and C III] 190.9, which seem likely to originate in the star since the nebula is of low excitation (Lutz and Carnochan 1979). The ultraviolet colour temperatures of the high-temperature stars agree with the He II Zanstra temperature. Some of the lower temperature stars however show a better agreement of the ultraviolet colour temperature with the H I Zanstra temperature, rather than with the higher He II Zanstra temperature (Pottasch *et al* 1978).

Some central stars are known to be double (e.g. the central star of NGC 1360, which is a single line spectroscopic binary (Mendez and Niemela 1979, Wehmeyer and Kohoutek 1979)) and the Zanstra method then fails completely since the line and continuum measurements may be determining the numbers of photons from different stars. The other main error of the method arises from the assumption of optical thickness. If it is not valid then some high-energy photons will escape and the temperature that is determined will be too low. Evidence that such an error may be occurring is found from the observation that temperatures obtained from He II are consistently higher than those from hydrogen (table 3.7). An independent method due to Khromov (1968) where the temperature is determined from a comparison of the sizes of the Stromgren spheres of He III and Ne IV, O V and N V also gives much higher temperatures (table 3.7).

Table 3.7. Temperatures of the central stars of planetary nebulae (Harmon and Seaton 1966, Heap 1977a, Khromov 1968, Lang 1978, Osterbrock 1974). T_1 – hydrogen Zanstra temperature, T_2 – He II Zanstra temperature, T_3 – Khromov temperature.

Nebula	T_1 (10^3 K)	T_1 (10^3 K)	T_3 (10^3 K)
NGC 1535	>37	74	
NGC 2392	84	68.5	160
NGC 2440	100	100	100
NGC 6210	43–50	70	
NGC 6818	138	200	140
NGC 6826	>40	44–69	
NGC 6886	90		160
NGC 6891	>28	52–55	
NGC 7662	100		130

The use of model atmospheres brings the temperatures into better agreement (table 3.8).

Central star temperatures can also be determined from the O III/O II line ratio in the nebula (using in particular the [O III] 500.7/([O II] 372.7 + [O II] 372.9) ratio), and are in broad agreement with the temperatures obtained by other methods (Koppen and Tarafdar 1978).

For those planetary nebulae whose distance is known (table 3.4), the absolute magnitude of the central star may be found, and these are listed in table 3.9. Their luminosities range from one tenth to a few hundred solar luminosities.

If the temperatures discussed earlier are effective temperatures, then it is straightforward to find the radii of the stars. However it is by no means clear that the temperatures may be assumed to be effective temperatures, and the radius estimates for the central stars therefore may contain large (and unknown) errors. With this caution in mind, some radii are listed in table 3.10.

Table 3.8. Zanstra temperatures of the central stars of planetary nebulae (Heap 1977a, Kaler 1976, Osterbrock 1974).

Nebula	Black body temperature (10^3 K)	Model atmosphere temperature (10^3 K)
NGC 40	32	32
NGC 1535	>37	94
NGC 2392	68	84–92
NGC 3242	93	109
NGC 3587	105	123
NGC 6210	43–58	45–92
NGC 6369	65	49
NGC 6543	42–66	39–82
NGC 6567	56	44
NGC 6572	60–61	46–77
NGC 6629	45	40
NGC 6790	66	49
NGC 6803	62	47
NGC 6807	60	46
NGC 6826	44–69	40–85
NGC 6833	49	41
NGC 6853	132	148
NGC 6891	28–55	42–76
NGC 7009	81	98
NGC 7662	100	118
IC 351	91	108
IC 418	34–43	33
IC 2149	37	34
IC 3568	61	47
IC 4406	73	53
IC 4593	42	38
IC 4634	55	44
IC 4732	64	48
IC 4846	61	47
IC 4997	49	42
BD + 30 3639	30	31
M 1–58	50	42
M 1–59	40	37
M 1–64	57	45
M 1–73	36	33
M 2–23	44	40
Me 1–1	65	49
Me 2–2	42	38
Hu 2–1	37	34
J 320	57	45
Vyl–1	52	42

Table 3.9. Absolute photographic magnitudes of the central stars of planetary nebulae.

Nebula	M_p	Nebula	M_p
NGC 246	+ 3.4	NGC 7009	+ 2.1
NGC 2392	− 0.4	NGC 7027	+ 3.7
NGC 3587	+ 5.0	NGC 7293	+ 7.6
NGC 3918	+ 3.6	NGC 7662	+ 1.9
NGC 6210	− 1.3	IC 418	− 1.1
NGC 6543	+ 0.3		
NGC 6572	− 0.4		
NGC 6720	+ 4.4		
NGC 6826	+ 0.5		
NGC 6853	+ 6.6		

Table 3.10. Radii of central stars of planetary nebulae.

Nebula	Radius ($R_\odot$)	Nebula	Radius ($R_\odot$)
NGC 2392	1.0	NGC 6826	0.7
NGC 3587	0.07	NGC 6853	0.003
NGC 6210	2.0	NGC 7009	0.3
NGC 6543	0.7	NGC 7662	0.3
NGC 6572	1.0	IC 418	2.0

The masses of the central stars are uncertain. On theoretical grounds, masses of 0.6 to $1.2 M_\odot$ are possible (Osterbrock 1974). One nebula, A 63, has an eclipsing binary, UU Sge, for its nucleus, and this leads to an estimate of $0.9 M_\odot$ for the primary and $0.7 M_\odot$ for the secondary (Bond *et al* 1978). Another central star, FG Sge has a 70 day oscillation in its light output (Whitney 1978). If this is attributed to a radial pulsation, then it suggests a mass of $0.41 M_\odot$.

3.10. Origin and Evolution of Planetary Nebulae

It is clear from the size and rate of expansion that the nebulae have a relatively short life – 20 000 years is sufficient for a nebula to expand from zero to a parsec in diameter at 25 km s^{-1}. At the same time their distribution suggests that they belong to intermediate population I, or old disc population, which would give an age of 10^9 years. The nebula can therefore only be a short phase in the evolution of the central star. The position of the central stars on the H–R diagram (figure 3.25) suggests that they are population II stars just prior to becoming white dwarfs, a conclusion supported by the small radii found for some of them (table 3.10). The youngest nebulae are found at position A in figure 3.25, while successively older ones are found at B and C.

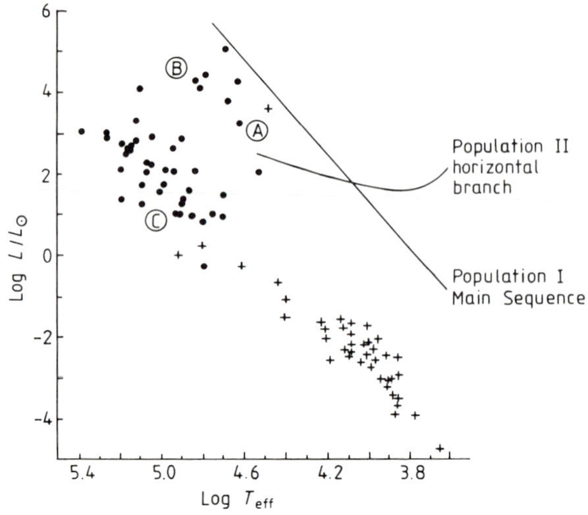

Figure 3.25. Position of the central stars of planetary nebulae on the H–R diagram. (From *Astrophysics of Gaseous Nebulae* by D E Osterbrock. W H Freeman and Co. Copyright © 1974.)

Possible candidates for proto-planetary nebulae are M 1–92, CRL 2789, MZ–3, M 2–9, CRL 618, and the Boomerang Nebula (Calvet and Cohen 1978, Taylor and Scarrott 1980). These are bi-polar nebulae, 10 to 40 arc seconds across, associated with strong, centred infrared sources, and with central stars whose temperature is around 30 000 K. The polarisation curve of the Boomerang Nebula (figure 3.26) suggests a model which consists of a dust disc centred on the star, with the grains aligned by a magnetic field in the nebula. The bi-polar lobes are produced by mass loss from the star (Taylor and Scarrott 1980). CRL 618 has a featureless 2 to 14 μm spectrum (Russell *et al* 1978). It has been detected at radio wavelengths from 5 to 15 GHz with intensities of 6 to 60 mJy, which is also consistent with its interpretation as a proto-planetary nebula (Wynn-Williams 1977). Such an interpretation however is not unique, and CRL 618 and M 1–92 have also been interpreted as pre-main sequence objects (Allen *et al* 1980).

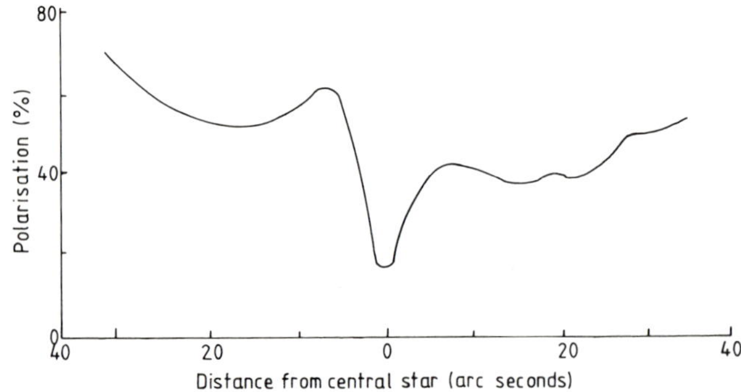

Figure 3.26. Polarisation along the major axis of the Boomerang Nebula. (Reproduced from Taylor and Scarrott 1980 by permission.

Other possibilities for proto-planetary nebulae include HM Sge, a red star which brightened by six magnitudes in 1975, and developed planetary nebula types of emission lines, infrared and radio spectra. Its spectrum is evolving to higher and higher levels of excitation with time. Expansion velocities of 35 to 1700 km s^{-1} are found, and its shell mass is estimated at $10^{-3} M_\odot$ (Ciatti *et al* 1977, 1979, Kwok and Purton 1979). V 1016 Cyg may be a similar system (Ciatti *et al* 1978, 1979, Fitzgerald and Pilavaki 1974). CRL 2688, a post-carbon star object with molecular radio emission lines also seems likely to evolve into a planetary nebula (Lo and Bechis 1976, Zuckermann *et al* 1976).

The masses of the central stars are only very poorly known (see earlier discussion). Calculations of evolutionary models suggest that the mass of the original star will divide along the lines of the ratios shown in table 3.11.

Table 3.11. Proportions of the mass in the shell and central star in planetary nebulae (Osterbrock 1974).

Original mass ($M_\odot$)	Mass of nebula ($M_\odot$)	Mass of central star ($M_\odot$)
0.8	0.2	0.6
1.5	0.7	0.8
3.0	1.8	1.2

The masses of the nebulae determined observationally (table 3.5) are low (less than $0.2 M_\odot$), suggesting a mass for the original star of $0.8 M_\odot$ or less. However since stars with masses less than about $0.9 M_\odot$ will not have evolved to the planetary nebula stage during the lifetime of the galaxy, the original mass must have been higher than this.

The number of planetary nebulae (10^4–4×10^4 (Cahn and Wyatt 1976, Jacoby 1980)) and their lifetime (less than 10^5 years) leads to an estimate of around 2×10^{10} for the number of ex-central stars of planetary nebulae now present in the galaxy. The estimated number of white dwarfs in the galaxy is 1.5×10^{10}, and the rates of formation of planetary nebulae and white dwarfs, together with the 'death' rate of 1 to 5 solar mass stars, are all currently comparable at about 2 to 4×10^{-3} kpc^{-3} a^{-1} (Cahn and Wyatt 1976, Weidemann 1977). Thus the nebulae and their central stars may almost certainly be seen on an evolutionary picture as white dwarf precursors. The velocities observed in the nebulae, however, are much lower than the escape velocities of the stars. In highly luminous red giants ($L \simeq 10^4 L_\odot$) the radiation pressure is sufficient to reduce considerably the effective gravity. It may be that planetary nebulae arise purely from radiation pressure overcoming gravity (i.e. a radiatively driven stellar wind – see chapter 1). It is more likely however that before this occurs, the total energy of the outer layers of the star (including ionisation energy) becomes positive. Pulsations will then be unstable and will increase in amplitude exponentially until the entire surface is lifted off the star, the process continuing until only the core (with little hydrogen) is left. In such a stage the star may resemble a Mira variable (Tuchman *et al* 1978, 1979). The pulsations will set in as soon as the total energy becomes positive, and so only a small velocity, perhaps 30 km s^{-1}, would be expected at infinity.

An alternative model has been proposed by Livio *et al* (1979) for planetary nebulae nuclei which are close binary systems. In their model the system is a red giant which fills its Roche lobe (see chapter 6 for a discussion of the Roche equipotential surfaces), and a main sequence star. The red giant loses mass to its companion, causing it to expand, fill its Roche lobe and then lose mass through L_2 (one of the outer Lagrangian points – see chapter 6). Eventually a steady state is reached with mass flowing through L_1 from the red giant and then out of the system through L_2. The ejected material then eventually forms the planetary nebula.

A third scenario has been proposed by Ferch and Salpeter (1975), and by Kwok *et al* (1978) in which a red giant loses mass through a stellar wind driven by radiation pressure on dust particles (see chapter 1), until it is stripped down to its core. An ultraviolet resonance line driven stellar wind then develops, which interacts with the dust shell and pushes it further out from the star to form the nebula. The emission line spectrum develops comparatively rapidly (as in HM Sge) when photoionisation overcomes recombination and attenuation by dust.

The spectra of the Wolf-Rayet and Of type central stars, by analogy with their higher mass counterparts, suggest expansion velocities of 10^3 km s^{-1} or more. Yet the nebulae even in these cases are expanding at only a few tens of kilometres per second. It is likely that the high-velocity region develops close to the star after the formation of the nebula, perhaps in a Ferch-Salpeter-Kwok model.

Symbiotic stars have been suggested as the immediate precursors to planetary nebulae, but there are too few of them to explain all the observed nebulae. Furthermore, a number of difficulties (particularly the general lack of cool companions to the central stars) militate against this hypothesis. Red giants and/or Mira variables are therefore almost certainly the major, if not the only, predecessors of planetary nebulae.

The amount of material returned by planetary nebulae to the interstellar medium is about $5M_\odot$ a^{-1}. The total amount returned by all stars is estimated to be about $30M_\odot$ a^{-1}. Planetary nebulae are thus the second most important contributors of material to the interstellar medium. (The most important are long period variables which contribute about $20M_\odot$ a^{-1}.) Their major significance lies in that their material has been appreciably processed in a stellar interior, and is the main cause of evolution of the composition of the interstellar medium. (Supernovae for example contribute only about 1% of the amount of material from planetary nebulae, although their material is much more thoroughly processed of course.)

4. *Cataclysmic Variables*

Many types of variables could be included under this heading. But we shall consider only four of the five types which are traditionally so labelled:

(i) Classical novae
(ii) Recurrent novae
(iii) Nova-like variables
(iv) Dwarf novae.

Supernovae (the fifth traditional type of cataclysmic variable) will not be considered in this work because a book far longer than this would need to be devoted solely to them to do them justice. (It has been said that there are two types of astronomy – astronomy of the Crab Nebula, and astronomy of everything else, and the former is the more important!)

The first three types listed above may be different varieties of a single type – the recurrent nova, the classical novae having periods so long that only one eruption has been observed, and the nova-like variables being long-period recurrent novae at minimum. The dwarf novae are sufficiently different to warrant a separate section.

4.1. **Recurrent Novae** (including classical novae and nova-like variables)

4.1.1. Light Curve

The light curve of a typical classical nova are shown in figure 4.1. Most of the characteristic features of a nova outburst are identifiable.

(*a*) The initial rise in brightness is extremely large and rapid – more than 11 magnitudes in two days for Nova Cyg 1978. The usual total change in brightness is 12 to 13 magnitudes, but may be higher (for example Nova Cyg 1975 which brightened by at least 19 magnitudes). The recurrent novae have smaller amplitudes (e.g. six magnitudes for T Pyx).

(*b*) A rapid fall from maximum. The time taken for the initial, more rapid part of the decline, defines the type of nova. Fast novae decline by three magnitudes in a few days, while slow novae take several months or even a year to fall by the same amount. The absolute magnitude at maximum is related to this initial rate of decline (figure 4.2). This relationship is very useful in deriving the distance of the nova (and also for finding the distances to galaxies for extra-galactic novae).

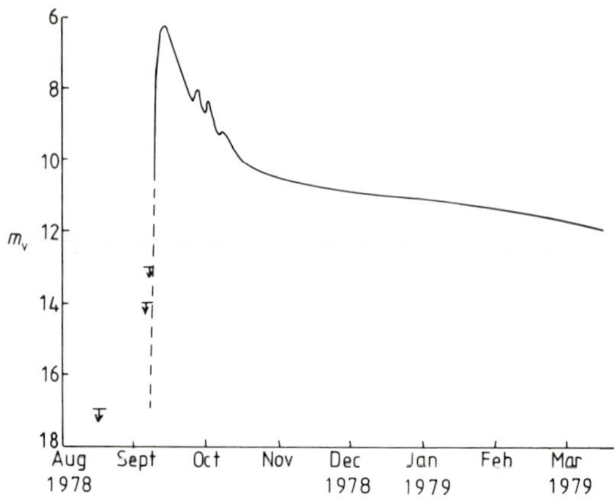

Figure 4.1. Light curve of Nova Cyg (1978). (Compiled from many sources.)

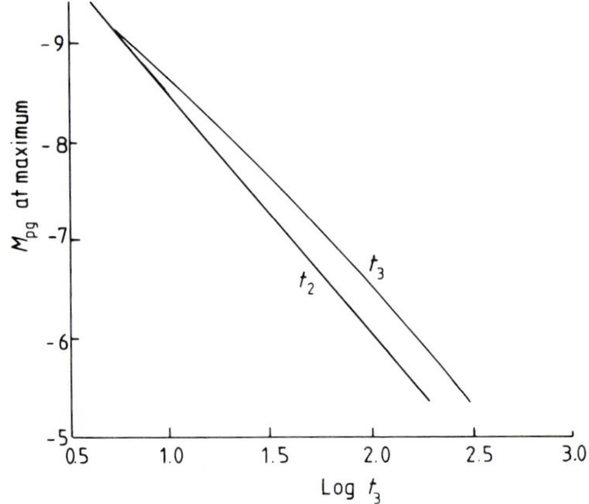

Figure 4.2. A plot of t_2 and t_3 (the times, in days, for a nova to decline by two and three magnitudes respectively from its maximum) against the absolute photographic magnitude at maximum.

(*c*) A slower rise over the last one or two magnitudes before maximum. In many novae the initial rapid rise may stop completely and even decline slightly before finally continuing to the maximum. (This is sometimes called the pre-maximum halt.)

(*d*) A series of fluctuations in brightness near the end of the rapid decline in brightness – known as the transition phase. These fluctuations may be much greater than in the case of Nova Cyg 1978, with the nova sometimes fading abruptly by several magnitudes for a short time.

(*e*) A slow, nearly exponential decrease in brightness. For recurrent novae this will return the object to its original brightness.

The ultraviolet part of the spectrum differs greatly from this behaviour. The ultraviolet flux continues to increase for some time after the optical maximum, so that their combined power is nearly constant (figure 4.3). The behaviour in the infrared is also quite different. The output continues to rise for 100 to 300 days after the optical maximum, until it reaches about 2×10^{31} W (Geisel *et al* 1970). Thereafter it decays exponentially with a time constant of about 150 days.

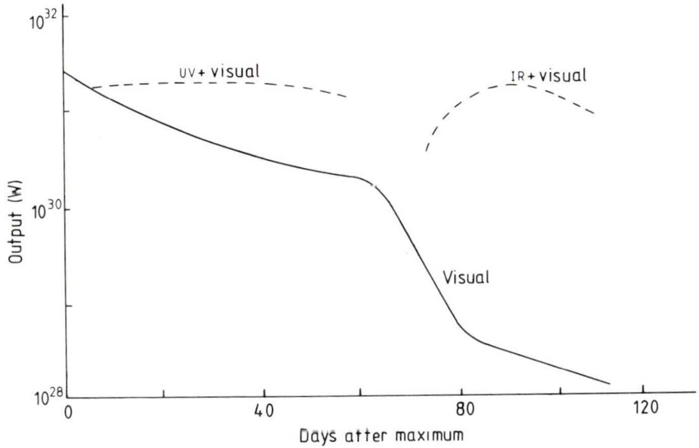

Figure 4.3. Visual, ultraviolet and infrared light curves for FH Ser (Nova Ser 1970). (From Gallagher and Starrfield 1978. Reproduced with permission from *Ann. Rev. Astron. Astrophys.*, **16**. © 1978 Annual Reviews Inc.)

The infrared emission seems likely to be due to the formation of dust in the outer parts of the ejecta. (Alternatively the dust may be in existence before the outburst, and be observed by reason of its rise in temperature (Bode and Evans 1980). Significant amounts of dust may form as early as five to ten days after the optical maximum (Phillips *et al* 1979). In Nova Cyg 1978 the dust shell remained optically thin, with a maximum of $\tau \simeq 0.1$ some 60 days after the optical maximum (Gerhz *et al* 1980b). DQ Her, and Nova Ser 1978 had optically thick shells, which in the latter reached a maximum 75 days after the outburst, when its mass was estimated to be $3.6 \times 10^{-7} M_{\odot}$. The size of the grains is about 0.3 μm (Gehrz *et al* 1980a). Gallagher (1977) has suggested a formula,

$$t_{\mathrm{d}} = \frac{320}{v} \left(\frac{L}{L_{\odot}} \right)^{1/2} \text{ days} \tag{4.1}$$

for the time of observable dust to form in an expanding nova shell (v is the expansion velocity in km s^{-1}. and L is the total luminosity during the constant post-maximum phase). However, in Nova Cyg 1975 thermal re-emission by dust did not develop until 300 days after the outburst. Prior to that the infrared emission was free–free thermal emission from a hot plasma. The plasma was initially optically thick, becoming optically thin later, and had a mass of about $10^{-4} M_{\odot}$ (Enmis *et al* 1977, Ferland and Wootten 1977). The amplitude of this nova was anomalously large so that this is probably not a typical development.

Emission features have been found at 2.2 and 4.8 μm in NQ Vul which are probably attributable to carbon monoxide (Ferland *et al* 1979, Ney and Hatfield 1978), while [Ne II] emission may explain the feature near 10 μm in Nova Cyg 1975 (Ferland and Shields 1978a). Nova Cyg 1975 has also been detected between 0.6 and 90 GHz with a typical intensity of 10 mJy, and a maximum of 260 mJy. The distribution is consistent with the emission arising as free–free radiation from an expanding cloud of ionised gas (Schwartz and Spencer 1977, Seaquist *et al* 1978). A similar model fits the radio emission of HR Del and FH Ser (Hjellming *et al* 1979).

The initial rate of decline in the visual is related to the total range of the nova outburst, Δm, by (Bertaud 1948)

$$\Delta m = 14.74^{\mathrm{m}} - 2.80^{\mathrm{m}} \log_{10} t_3 \tag{4.2}$$

(where t_3 is the time in days for the nova to decline by three magnitudes from maximum) for classical novae. The recurrent novae have much lower amplitudes than would be predicted by this formula (figure 4.4). However, all recurrent novae known at present appear to have cool giants as companions. Thus the T CrB system contains an M3 III star, V 1017 Sgr a G5 III star and RS Oph an M6 star (Warner 1975). (Another possibility is WZ Sge, but this is more likely to be a very long-period dwarf nova, or a dwarf nova in an extended standstill.) Their magnitudes at minimum are therefore those of their companions and not of the accretion ring of material around the white dwarf (models of novae are discussed further, later in this chapter). There is no reason to doubt that recurrent novae would have amplitudes comparable with classical novae if the brightness of the giant companion were removed.

The timescales given above are appropriate for fast novae. Slow novae have similar light curves with all the rates of change appropriately lengthened (figure 4.5).

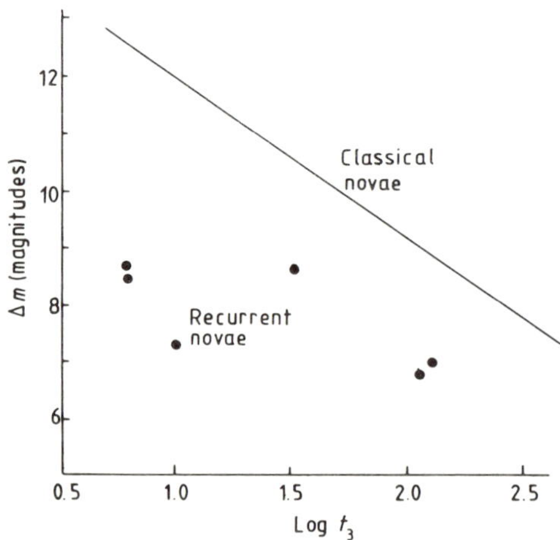

Figure 4.4. Variation of the range of nova outbursts with t_3.

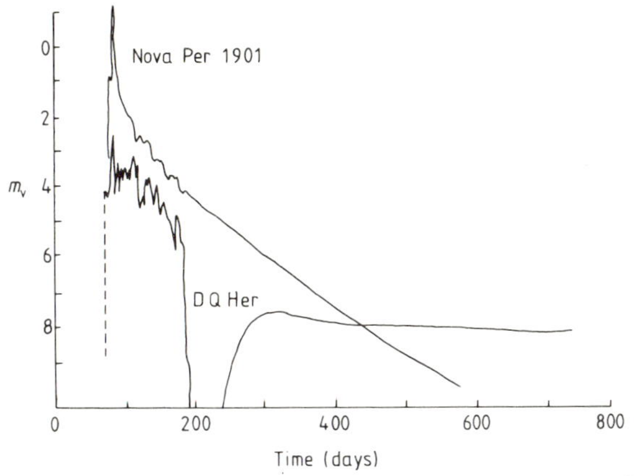

Figure 4.5. Light curves of fast (Nova Per 1901) and slow (DQ Her) novae (Campbell 1919, Grouillier 1935).

A possible identification of a transient x-ray source in Perseus, A 0327 + 43, is the old nova, GK Per, The x-ray source was observed in June 1978 and at the same time GK Per had brightened optically by one magnitude. The x-ray source had a peak intensity of 2×10^{-13} W m^{-2} over the range 2 to 10 keV (King *et al* 1979). Nova Oph 1977 may also be identifiable with the bright x-ray transient H 1705 − 25 (Griffiths *et al* 1978, Watson *et al* 1978a). In AE Aqr the 0.1 to 4.0 keV emission has a 33 second period which matches the optical flickering (Patterson *et al* 1980). Medium energy x-rays have been found from Nova Mon 1975 and A 1524 − 61 during their outbursts, with the x-ray emission peaking at the same time as the optimal maximum (Murdin *et al* 1977). One nova-like variable, MV Lyr also has soft x-ray emission from its vicinity, but no positive identification has been made (Mason *et al* 1979). On the other hand both Nova Cyg 1975 and Nova Vul 1976 were searched for in the x-ray region, but were not found (Cruise 1977).

The total amount of energy radiated by most novae during an outburst is 10^{38} to 10^{39} J (for comparison, the Sun radiates about 10^{34} J in a year). The repetitive eruptions of recurrent novae (periods 10 to 80 years) could lead to their total long-term energy loss significantly exceeding that of classical novae if these only erupted once. The similarities between the light curves of recurrent and classical novae are so marked, however, that it is not possible to identify a nova as recurrent from its first observed outburst. Only the observation of subsequent outbursts will do so. Their behaviour between outbursts is also very similar to that of post-classical novae. Pre-eruption light curves of eleven novae have been compiled by Robinson (1975), and are indistinguishable from their post-eruption light curves. Thus it is likely that a majority, if not all, novae are recurrent. The observed recurrent novae are just those with the shortest periods. It may be, however, that the presence of the giant companions in the known recurrent novae, is the cause of their repetition. The only other novae with giant companions, RR Tel and RT Ser, are anomalous slow novae

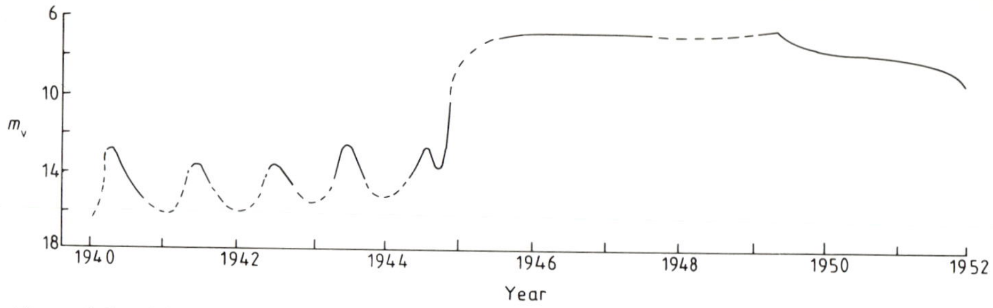

Figure 4.6. Light curve of RR Tel (Seaquist and Palimaka 1977).

(figure 4.6) so that the interpretation of classical novae as long-period recurrent novae cannot be regarded as finally settled.

At minimum, a nova may be constant (e.g. T Aur, DI Lac), or vary slowly (e.g. T CrB), or rapidly (e.g. V841 Oph, GK Per), by one or two magnitudes (figure 4.7). The mean absolute visual magnitude at minimum of those novae bright enough to be observed then is $+4^m$ (Payne-Gaposhkin 1957).

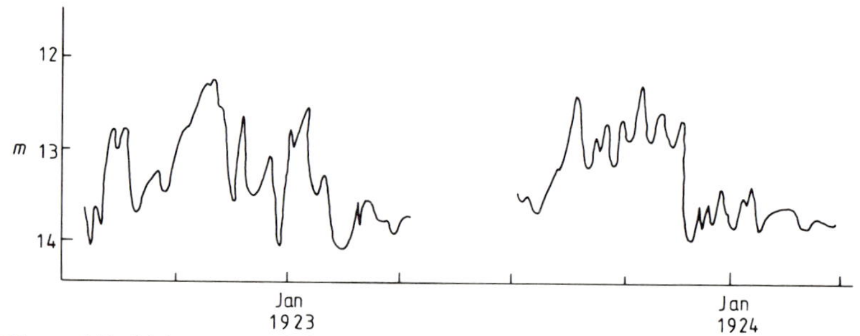

Figure 4.7. Light curve of GK Per (Nova Per 1901) at minimum. (Reproduced from Steavenson 1923, 1924 by permission.)

Nova Cyg 1975 has shown a 3.3 to 3.4 hour modulation to its brightness since its outburst (Kleine and Kohoutek 1979). This is possibly attributable to a spiral density wave in the stellar wind, which originates in the binary motion of the central stars, the wave persisting far enough out into the envelope to produce visible effects (Fabian and Pringle 1977, Hutchings and McCall 1977). A possibly similar variation has been found in Nova Cyg 1978 where the period is 10 h 32 min and the amplitude 0.15 magnitudes (Campolonghi *et al* 1980).

Very rapid variations in brightness are common in dwarf novae (see later in this chapter), but have only been observed in a few novae. DQ Her in particular has maintained a sinusoidal 0.04^m oscillation with a period of 71 seconds for many years (Herbst *et al* 1974). The amplitude is greatest just after the eclipse in this system, and the oscillations show a steady shift in their phase which totals 360° throughout the eclipse (Warner 1975). There is also an increase in phase with increasing wavelength across the He II 468.6 emission line (Chanan *et al* 1978), and there is a variation in polarisation (both linear and circular) which is related to the magnitude changes, but with twice the period (Kemp *et al* 1974, Swedlund *et al* 1974). The 71 second period is decreasing very slowly with,

$$\dot{P} = -6.04 \times 10^{-11} \pm 0.18 \times 10^{-11}. \tag{4.3}$$

(This is almost comparable with the stability of the periods of pulsars, for which $\dot{P}$ ranges from 10^{-16} to 4×10^{-13}.) Amongst the other classical and recurrent novae, and the nova-like variables, a few have shown similar stable fluctuations, usually only for short periods of time (table 4.1).

Table 4.1. Rapid coherent variations in novae (Giuricin *et al* 1979, Kemp *et al* 1977), Mardirossian *et al* 1980, Nather and Robinson 1974, Patterson 1979a, b, Patterson *et al* 1980, Robinson 1976a, Warner 1975).

Name	Period (s)	Amplitude (^{m})	Type
AE Aqr	33.076737	0.01	Nova-like variable
TT Ari	20, 40	0.1	Nova-like variable
AM CVn	112.8, 121.0		Nova-like variable
Nova Cyg 1975	~100		Classical nova
Nova Cyg 1978	0.01, 5.3		Classical nova
DQ Her	71.065	0.04	Classical nova
V 533 Her	63.63309	0.01	Classical nova
RR Pic	26.2, 28.7, 36.9		Classical nova
UX UMa	28.54, 30.03	0.002	Nova-like variable
CD −42° 14462	18.50, 29.08, 30.15, 32.0	0.003	Nova-like variable

Most novae occurring in the Milky Way Galaxy are obscured by gas and dust and so are not observed. From observations of nearby galaxies it is estimated that between 10 and 40 occur each year in an average galaxy. This is further evidence in favour of all novae being recurrent. For, if this were not the case, and if, say, 25 novae per year had occurred in the Milky Way Galaxy throughout its life, then about 2×10^{11} stars would have undergone nova explosions. Since this number is comparable with the total number of stars in the galaxy, and since the majority of those stars have not evolved off the main sequence, we may reject such an hypothesis. Bath and Shaviv (1978) obtain an interval of 10^5 years between the outbursts of an average classical nova on the basis of a similar argument to that given above. A typical nova probably undergoes 160 to 660 outbursts in total (Ford 1977).

4.1.2. Spectrum

Very little is known of the spectra of pre-classical novae. A few low dispersion objective prism spectrograms exist. In HR Del there is a featureless blue continuum with no lines or Balmer discontinuity, and the energy distribution is that of an O or early B type star (Stephenson 1967). This appearance is similar to that of some other novae at post-nova minimum when taken at the same dispersion. Also, since it seems likely that all novae are recurrent, we may equate the pre- and post-nova spectra.

Details of the spectra at minimum of a large number of novae have been listed by Warner (1975) and the main features (not all of which are present in all novae) are summarised below:

Hydrogen:	Broad and in emission, varying from weak to strong. Can be very narrow, and in one case has a broad underlying absorption.
Helium:	Lines usually due to ionised helium. Broad emission lines, sometimes double. Occasionally weak He I emission.
Calcium:	Occasionally weak ionised emission.
Nitrogen } Carbon }	Doubly ionised emission occasionally present at 465 nm.
Others:	The recurrent novae frequently contain lines due to their giant companions. In several nova, nebular lines appear.

The nova-like variables (also known as rapid blue variables) have basically similar spectra. Absorption lines with central reversals are more common, however, and lines of singly ionised iron and oxygen may appear. The spectral types of old novae place them below the nuclei of planetary nebulae on the H–R diagram (figure 4.8).

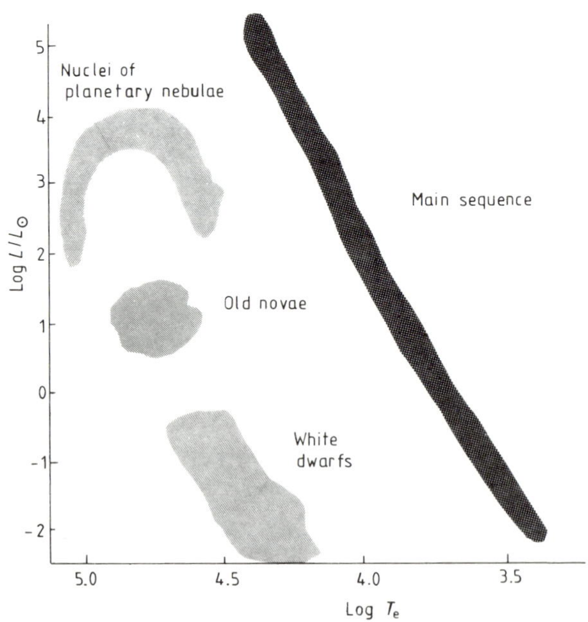

Figure 4.8. Position of old novae on the H–R diagram.

The spectra vary on all timescales. Of particular importance are the periodic velocity variations which imply that the nova is a binary. The spectrum may contain the lines of both stars, or one star and/or be an eclipsing system. In a few cases the lines of two stars are seen without any velocity variations, when the line of sight is presumably perpendicular to the orbital plane. The periods are short – 18 minutes to $16\frac{1}{2}$ hours – with one exception. The exception is T CrB whose period is 228 days. The two stars are usually a white dwarf with a relatively cool normal or giant star (referred to as the companion in this chapter). The companion fills the Roche lobe (for periods less than 10 hours, even dwarf stars will fill their Roche lobes, see chapter 6), and material spills from it through the inner Lagrangian point towards the white dwarf. It is generally believed that all nova are mass exchange binary systems.

The lines of the companion (when visible) are 180° out of phase with the emission lines. The emission region is therefore associated with the white dwarf. The nebular emission lines do not show the velocity variations and the region producing them must be detached from the binary.

During an outburst the spectrum undergoes changes whose magnitude is concomitant with the magnitude variations. We may identify five stages:

(*a*) Pre-maximum
(*b*) Maximum (or principal)
(*c*) Enhanced
(*d*) Orion
(*e*) Nebular.

Only the main features of these stages are summarised here. Payne-Gaposhkin (1957) gives an exhaustive treatment of the development of the optical spectrum and should be consulted for further details.

(*a*) *Pre-maximum spectrum.* The lines are mostly absorption lines, displaced by up to 1000 km s^{-1} to shorter wavelengths, and corresponding to spectral types B and A (and in one case F (McLaughlin 1960)). The lines of carbon, nitrogen and oxygen are enhanced, almost certainly due to enhanced abundance of these elements. In Nova Cep 1971 the O/H ratio was 10 times the cosmic value, and the N/H ratio nearly 1000 times larger (de Freitas-Pacheco 1977). Similarly carbon is overabundant in DQ Her by about a factor of 70 (Ferland and Truran 1980), and carbon, nitrogen, oxygen and neon are enhanced by factors of 20 to 100 in Nova Cyg 1975 (Ferland and Shields 1978b). Helium may also be over-abundant compared with the cosmic value, perhaps by as much as a factor of three (Ferland 1979a, de Freitas-Pacheco 1977, 1979, Gallagher *et al* 1980). Isotope abundances also seem to be peculiar. In DQ Her the $^{12}C/^{13}C$ ratio was greater than about 1.5, and the $^{14}N/^{15}N$ ratio was greater than about 2 (Sneden and Lambert 1975). The ultraviolet spectrum of only one classical nova, Nova Cyg 1978, has been caught before maximum, and this is shown in figure 4.9. The flux falls very steeply to shorter wavelengths, and the main features are due to neutral, singly and doubly

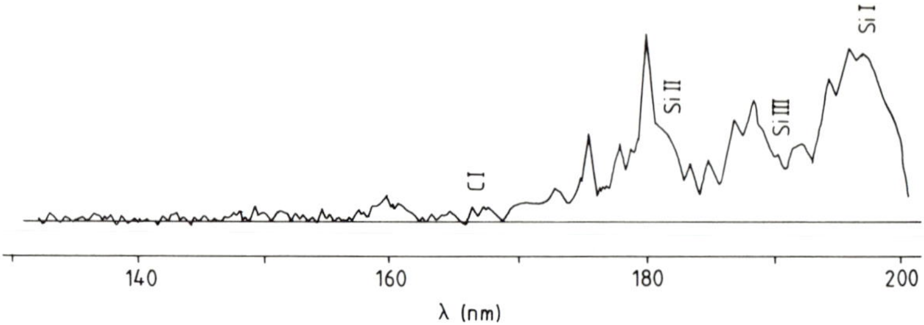

Figure 4.9. Pre-maximum ultraviolet spectrum of Nova Cyg (1978).

ionised metals. The pre-maximum phase is very short, lasting at most for only one or two days.

(*b*) *Maximum (or principal) spectrum.* The spectrum at this time approximates that of an A or F type supergiant, again displaced to shorter wavelengths by up to 1000 km s^{-1}. As the luminosity diminishes a second set of absorption lines appears with a displacement to shorter wavelengths of up to 1500 km s^{-1}, which gradually replaces the original spectrum. Almost at the same time, emission components appear on the long-wavelength edge of some lines (particularly those due to H, Ca II, Na I and Fe II), producing P Cygni type line profiles. The continuum diminishes rapidly in intensity, and the emission components and emission lines of [O I] and [N II] start to dominate the spectrum. In the ultraviolet Mg II and Fe II lines appear (figure 4.10).

The short wavelength spectrum (less than 200 nm) is very faint. The recurrent nova, WZ Sge, after 33 years of quiescence had its most recent outburst starting on

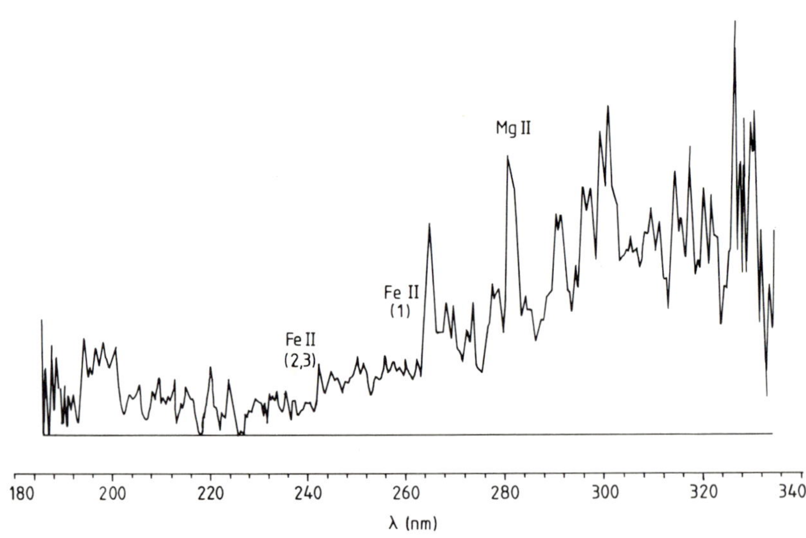

Figure 4.10. Spectrum of Nova Cyg (1978) seven days after outburst.

30 November 1978. It is not a typical recurrent nova and may perhaps better be considered as a long-period dwarf nova (Warner 1975), or it may belong to a new sub-class of the novae with the similar systems UZ Boo, WX Cet (Bailey 1979), and perhaps EX Hya (Paczynski and Rudak 1980). It was observed extensively in the ultraviolet by the international ultraviolet explorer (IUE) satellite between 5 and 25 days after the beginning of its outburst (Fabian *et al* 1980). The spectra were extremely variable initially, with both the line spectrum and the continuum changing. The main emission feature is the C IV 155 line (figure 4.11), and this varied on timescales of only a few minutes. Three weeks after the outburst, the spectrum was much more stable and the continuum strength had decreased. The total flux may still have been unchanged since the decrease occurred mainly at the longer wavelengths and could be due to flux redistribution to shorter wavelengths. Further optical emission lines due to N III, He II, [O III], [Ne III] and [Fe II] appear as the absorption spectrum disappears. The spectrum slowly merges into the nebular phase. (Note that tracings of spectra apparently show the emission lines increasing in intensity. In fact they are weakening all the time, but the continuum is weakening even faster. Since increased exposures are naturally given by astronomers who want well exposed spectra, the emission lines appear to be strengthening.)

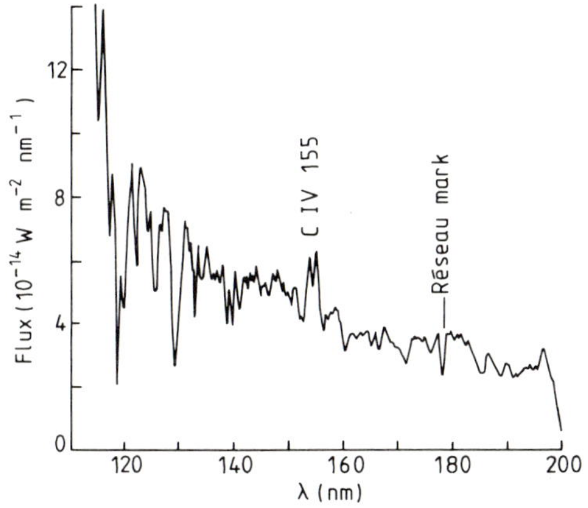

Figure 4.11. Ultraviolet spectrum of WZ Sge on 14 December 1978. (Reproduced from Fabian *et al* 1980 by permission.)

(*c*) *Enhanced spectrum.* This appears soon after optical maximum, while the maximum spectrum is still prominent. The lines are displaced to shorter wavelengths by about twice the velocity of the displacement of the maximum spectrum. As this spectrum develops, the initial wide, hazy absorptions due to H, Ca II, Mg II, Fe II, Na I and O I (and also possibly Ti II and Cr II) intensify and become narrower. Faint emission may be present as part of P Cygni line profiles. The lines then fade, and after a few days (less than or equal to t_3) they disappear.

Both the enhanced and maximum spectra may have several components to their lines, representing widely differing velocities.

(*d*) *Orion spectrum.* This appears as the enhanced spectrum reaches its maximum development and continues through to the transition stage of the light curve. The lines are absorption lines of H, N II and O II with velocity displacements at least as great as those of the enhanced spectrum. Hazy emission lines occur, and there is a broad emission near 465 nm which is probably attributable to N III and C III. Fast novae have N III 409.7 and 410.3 absorptions at this stage, but they are very weak or absent in slow novae (McLaughlin 1960).

(*e*) *Nebular spectrum.* The emission lines dominate the spectrum, and forbidden and high excitation lines become more and more important. The spectrum closely resembles that of a planetary nebula (chapter 3). In the ultraviolet, intercombination lines of O III], O IV], N III] and C III] are important as well as lines of N V, C IV, [Fe VII], [Fe X], [Fe XI] and [S VIII] (figure 4.12). The very high levels of ionisation are considered to arise in a gas shocked to a temperature of 10^6 K by supersonic chaotic motions (Shields and Ferland 1978). In the infrared, He I, O I and [O II] lines appear. The strong O I 844.6 emission (figure 4.13) arises through selective fluorescence with the Lyβ line (Strittmatter *et al* 1977) (see also chapters 1 and 5).

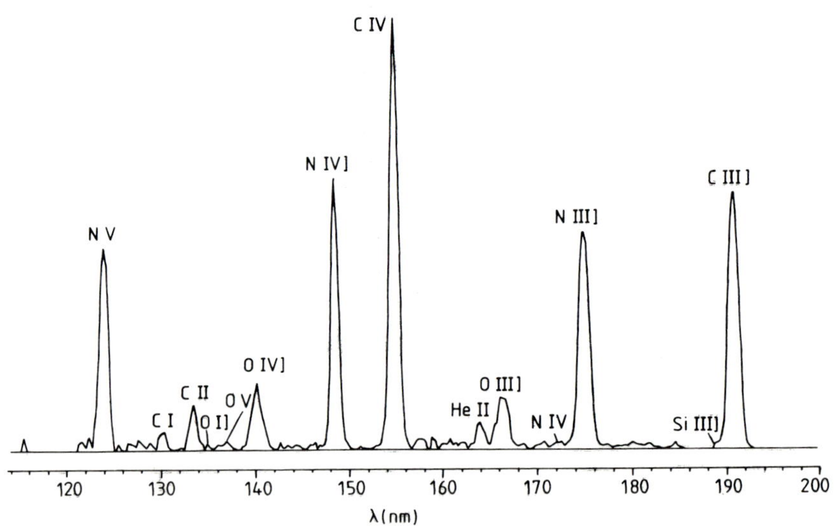

Figure 4.12. Ultraviolet spectrum of Nova Cyg (1978) in the nebular stage.

The spectrum of the recurrent nova, T Pyx, shows a similar development to that of the classical novae outlined above. But other recurrent novae differ from this standard pattern. T CrB for example exhibited lines of higher excitation, and did not develop enhanced, Orion, or nebular stages.

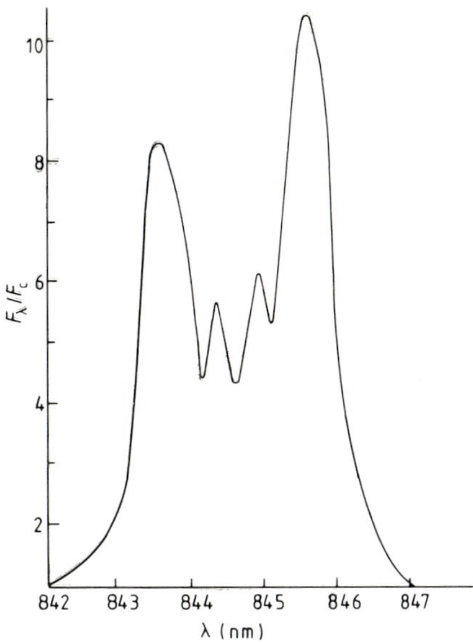

Figure 4.13. O I 844.6 in Nova Del (1967) on the 12 May 1968. (Reproduced from Andrillat and Houziaux 1969 by permission.)

4.1.3. *Masses*

Only one cataclysmic variable, the dwarf nova EM Cyg, is a double-line eclipsing spectroscopic binary, and so has a reliable mass estimate (see later in this chapter). For one recurrent nova, T CrB, the inclination is variously estimated to lie between 68° and 90° leading to masses 1.6 ± 0.4 to $2.1 \pm 0.5\ M_\odot$ (Warner 1975). However, the radial velocities are low in this system and may be distorted by the flow of material in the accretion disc. The listed errors are therefore only lower limits to the possible errors.

The masses of most novae can only be estimated from the assumption that the companion fills its Roche lobe. If q is the mass ratio (M_2/M_1), then we may write Kepler's third law as (after Warner 1975),

$$M_2 = \frac{4\pi^2}{G} \frac{q}{1+q} \frac{a^3}{P^2} \tag{4.4}$$

where M_2 is the mass of the companion (kg), a is the semi-major axis of the system (m) and P is the orbital period (s). Which, with R_2 for the radius of the companion, becomes,

$$R_2^3 M_2^{-1} = \frac{G}{4\pi^2} \frac{1+q}{q} \left(\frac{R_2}{a}\right)^3 P^2. \tag{4.5}$$

Now Paczynski (1971) gives the mean radius of a Roche lobe as

$$\frac{R_2}{a} \simeq \begin{cases} 0.38 + 0.2 \log_{10} q & 0.5 < q < 20 \quad (4.6) \\ 0.462 \left(\dfrac{q}{1+q}\right)^{1/3} & 0 < q < 0.5, \quad (4.7) \end{cases}$$

so that we may write:

$$R_2^3 M_2^{-1} = 1.7 \times 10^{-12} P^2 f(q), \tag{4.8}$$

where

$$f(q) = \frac{1+q}{q} \left(\frac{R_2}{a}\right)^3, \tag{4.9}$$

and is plotted in figure 4.14 for companions which fill their Roche lobes. It is a very insensitive function of q. Thus for any given M_2/R_2 relationship, we may determine the mass of the companion even if q is known only very inaccurately. For accurate values of q, the mass of the white dwarf may also be found. The masses obtained by this method are listed in table 4.2.

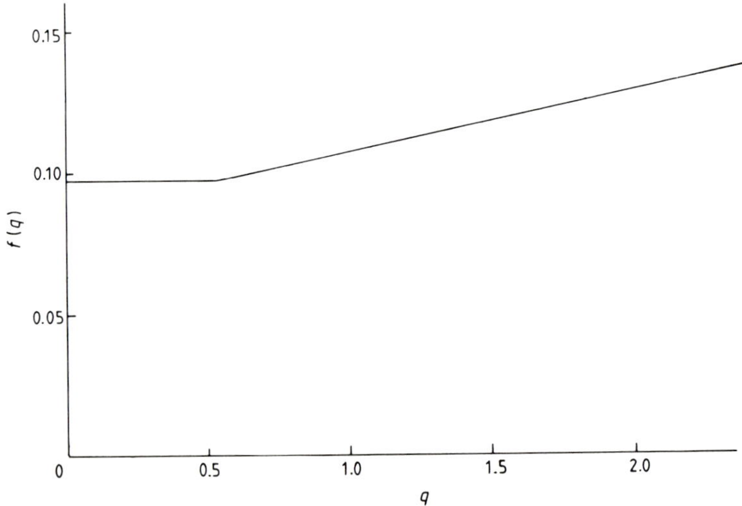

Figure 4.14. Variation of $f(q)$ with q.

The mass of material which is expelled by the nova is even more poorly known. Estimates using the methods based on line ratios developed for planetary nebulae (chapter 3) may be used, but have large errors. Thus the mass of the ejecta from HR Del is variously estimated between 1.8×10^{26} and 2.9×10^{27} kg. Radio emission from FH Ser (Seaquist and Palimaka 1977) gives a shell mass of 8×10^{25} kg, while modelling of the energy distribution suggests about 10^{25} kg. Hβ intensities in nova Cep 1971 suggest 2×10^{26} kg, and the Balmer series of nova Cyg 1975, 8×10^{25} kg

Table 4.2. Masses of novae (Bianchini 1980, Hutchings 1979a, Hutchings *et al* 1979, Ritter 1976, Warner 1975, Young and Schneider 1980).

Name	$M_2/M_\odot$	$M_1/M_\odot$	Type
V 603 Aql	0.40	0.87	Classical nova
AE Aqr	1.18	1.25–1.31	Nova-like variable
TT Ari	0.38	0.79	Nova-like variable
T Aur	0.63–0.68		Classical nova
HR Del	0.5	1.0	Slow nova
DQ Her	0.32–0.60	0.35–1.0	Classical nova
RR Pic	0.44		Classical nova
VV Pup	0.16		Nova-like variable
VZ Scl	0.44	0.32	Nova-like variable
RW Tri	0.71		Nova-like variable
UX UMa	0.61		Nova-like variable

(de Freitas-Pacheco 1977, Sanyal and Willson 1980). Overall a mass for the ejecta within a magnitude of 10^{26} kg ($5 \times 10^{-5} M_\odot$) seems likely. With ejection velocities of 1000 km s^{-1}, the kinetic energy involved in an outburst is therefore about 10^{36} J – 1% or less of the total energy of the outburst.

4.2. Dwarf Novae (U Gem, Z Cam type stars)

Although dwarf novae are sufficiently different from classical and recurrent novae to warrant a separate section, they do have many characteristics in common with the other novae. Therefore only the main distinguishing features of dwarf novae will be covered in this section.

4.2.1. *Light Curve*

The light of the system increases suddenly and rapidly in a nova-like fashion. The rise in brightness, however, is only two to six magnitudes, and after a day or so at maximum, it declines almost linearly back to the pre-nova level, the fall in brightness taking two to ten days to occur. After an interval which is usually between 20 and 50 days, but which can range between 10 and 400 days, the outburst repeats. Usually each outburst for a given dwarf nova follows a similar pattern to its previous outbursts. The interval between outbursts varies, but is usually sufficiently constant for a characteristic timescale to be usefully assigned to any given star. A typical schematic light curve is shown in figure 4.15. The majority of light curves for dwarf novae resemble this pattern, and such stars are sometimes given the sub-class, U Geminorum type stars.

Another sub-class, the Z Camelopardalis stars, comprise about 10% of the dwarf novae (Payne-Gaposhkin 1977), and are characterised by 'standstills' in their light

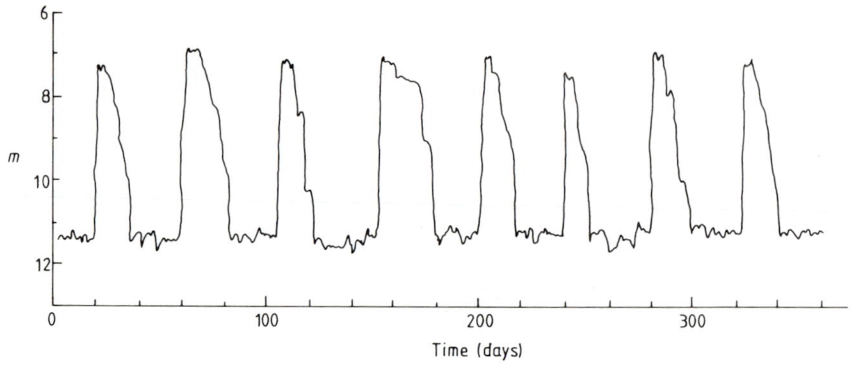

Figure 4.15. Schematic light curve of a dwarf nova.

curves. These almost always occur during the decline from an outburst (figure 4.16), usually at the same luminosity on each occasion. The light output remains nearly constant for a period, usually of only a few days, but the standstill can last for several years on rare occasions. The typical timescales for Z Cam type stars are shorter than average for dwarf novae, normally being less than 50 days.

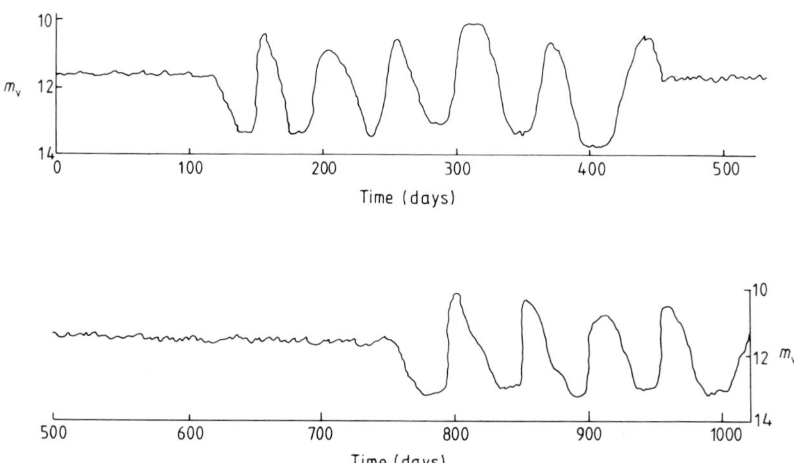

Figure 4.16. Schematic light curve of a Z Cam type of dwarf nova.

WZ Sge is a peculiar recurrent nova with a period of about 30 years which, as discussed earlier, may be better interpreted as a dwarf nova in a very prolonged standstill, or a very long-period U Gem star (Warner 1975). It may form the prototype of a new sub-class of the dwarf novae along with the similar systems, UZ Boo, WX Cet (Bailey 1979), and perhaps EX Hya (Fabian *et al* 1980) and UX UMa (Robinson 1976a).

Some dwarf novae (sometimes called the SU UMa type novae (Vogt 1980)) periodically undergo an extended outburst, called a supermaximum. They remain at or near their maximum magnitude for several times their normal duration (figure 4.17). The intervals between supermaxima are several hundred days (three to ten times the normal timescale between outbursts), and they occur more regularly than the normal maxima. Dwarf novae which have supermaxima are among those with the shortest periods, but there is no overlap with the Z Cam stars. The companions in this class of novae all seem to be M type main sequence stars (Vogt 1980). The mass exchange rate is higher during a supermaximum, when compared with a normal maximum, by about a factor of four (Mayo *et al* 1980).

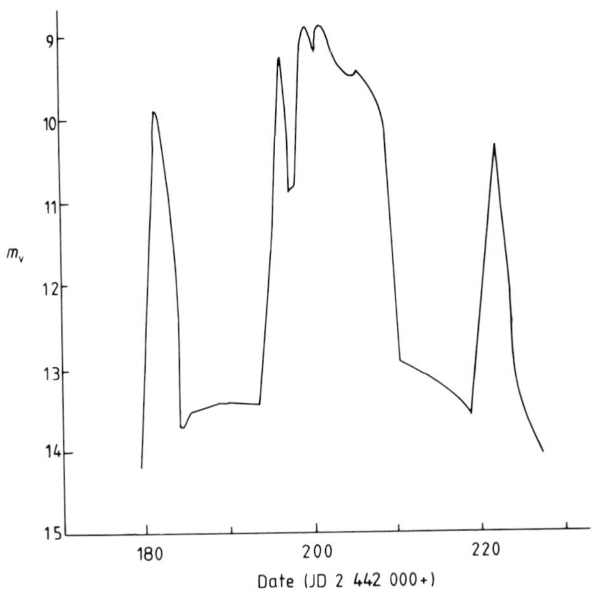

Figure 4.17. Normal and supermaxima in VW Hya. (Reproduced from Warner 1975 by permission.)

The longest period dwarf novae are sometimes classed as UV Per type stars. Finally CD $-42°$ 14462, V 751 Cyg, Vy Scl and perhaps UX UMa, which are sometimes classed as nova-like variables (or a WZ Sge type star for UX UMa) may perhaps be better interpreted as dwarf novae in an extended or permanent state of outburst (Bond *et al* 1978, Robinson *et al* 1974). A large number of light curves of individual dwarf novae are given by Glasby (1970).

On a two-colour diagram, the dwarf nova follows a loop during its outburst (Bailey 1980) (figure 4.18). This suggests that the radiation comes from the outer, cooler, parts of the accretion disc initially. Later in an outburst the radiation originates lower down in the hotter, inner, parts of the disc. (See later in this chapter for a discussion of models of novae.)

Two further features are found in the light curves: flickering and binary variations. The flickering is similar to that of novae (figure 4.19) although more common, and details are listed in table 4.3. Usually the flickering varies in period

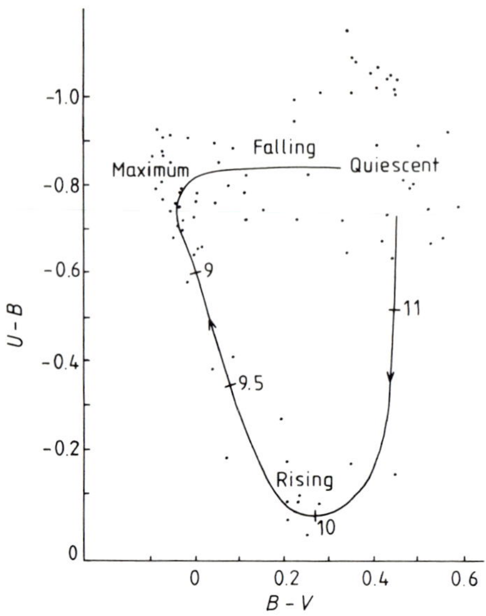

Figure 4.18. Two-colour diagram of SS Cyg during an outburst. (Reproduced from Bailey 1980 by permission.)

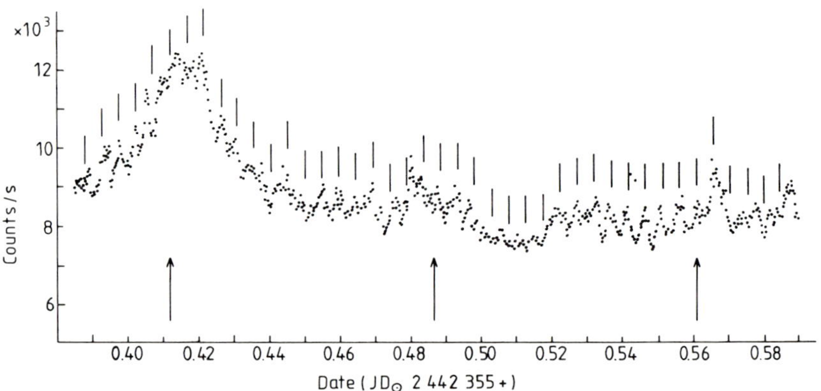

Figure 4.19. Light curve of VW Hyi on 2 November 1974. Arrows indicate expected humps. Vertical lines indicate the 413 s modulation. (Reproduced from Warner and Brickhill 1978 by permission.)

(figure 4.20), sometimes with a steady drift, at other times more irregularly, and perhaps with many periods present. In V 436 Cen the period correlates with the brightness, being longest when the system is brightest (Warner and Brickhill 1978) (figure 4.21). In AH Her the glickering disappears near the maximum of an outburst (Hildebrand *et al* 1980). It is stable to better than one part in 10^5 over many months for WZ Sge (Patterson 1980), but retains coherence for only a few periods and is only found during an outburst in U Gem, SS Cyg, KT Per, RU Peg and VW Hyi (Robinson and Nather 1979).

Table 4.3. Rapid coherent variations in dwarf novae (Haefner *et al* 1977, 1979, Hildebrand *et al* 1980, Horne and Gomer 1980, Moffett and Barnes 1974, Nevo and Sadeh 1976, 1978, Patterson 1980, Patterson *et al* 1978, 1977, Robinson 1976a, Robinson *et al* 1978, Stiening *et al* 1979, Szkody 1976, Warner 1975, Warner and Brickhill 1978).

Name	Period (s)	Amplitude (^{m})	Type
RX And	35.7	0.003	Z Cam
Z Cam	16.01, 16.92, 18.83	0.001	Z Cam
V 436 Cen	20.0, 20.5	0.002	U Gem
Z Cha	27.67	0.003	U Gem
SY Cnc	20.3–33.5	0.003	Z Cam
YZ Cnc	26.99	0.0018	U Gem
EM Cyg	16.6		Z Cam
SS Cyg	8.23–8.50, 9.735	0.0005	U Gem
AH Her	24.1–25.78, 31.55–34.8	0.003	Z Cam
VW Hyi	28.0–35.0, 88.0, 413.0	0.02	U Gem
CN Ori	24.27–25.00, 33.0	0.005	Z Cam
RU Peg	11.6, 48.0–55.0	0.003	U Gem
KT Per	26.730–26.825, 22.5–29.5	0.006	Z Cam
WZ Sge	27.87, 28.98		WZ Sge

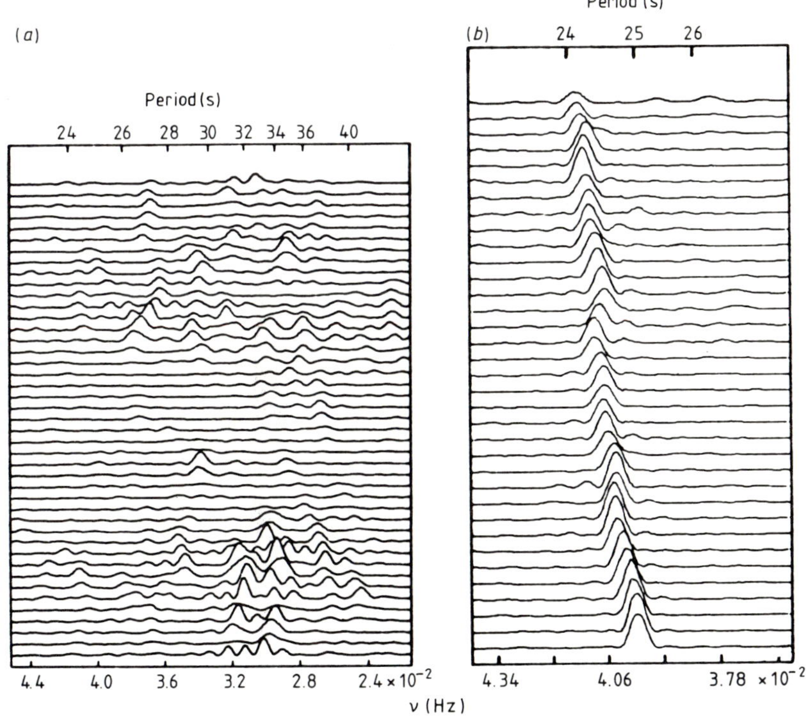

Figure 4.20. Periodograms of (*a*) VW Hyi on 6 December 1973 and (*b*) CN Ori on 12 February 1972. (Reproduced from Warner and Brickhill 1978 by permission.)

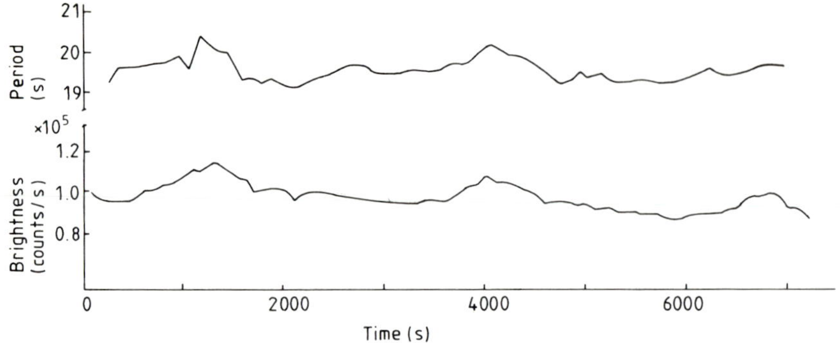

Figure 4.21. Brightness and flickering period correlations in V436 Cen on 11 June 1975. (Reproduced from Warner and Brickhill 1978 by permission.)

Most dwarf novae are observed to be binary systems, and all are believed to be binaries (table 4.4). Many of them exhibit light curves which reflect this binary nature and whose main features are shown in figure 4.22 (not all the features are present in all the systems). There is a hump in the light curve from about phase 0.6 to 0.1, which contains the eclipse, if any. The flickering is of larger amplitude during the hump (A), than normally (C), and it disappears entirely during eclipse (B). During an outburst the eclipse minima of some dwarf novae (e.g. U Gem) become very much shallower and may almost disappear, while in others (Z Cha) the shape of the light curve at eclipse minimum becomes more rounded in an outburst, but otherwise is little changed (Bateson 1978, Warner 1974).

Table 4.4. Orbital periods of dwarf nova binaries (Bailey and Ward 1981, Cowley *et al* 1980, Kiplinger 1979, Patterson *et al* 1979, Robinson *et al* 1978, Vogt and Breysacher 1980, Warner 1975, Warner and Thackeray 1975).

Name	Period (H min)	Type
RX And	5 05	Z Cam
SS Aur	4 20	U Gem
Z Cam	6 56	Z Cam
OY Car	1 31	
BV Cen	14 38	U Gem
V 436 Cen	1 32	U Gem
WW Cet	3 50	Z Cam
Z Cha	1 47	U Gem
AM CVn	0 17.5 (increasing)	U Gem
EM Cyg	7 00	Z Cam
SS Cyg	6 36–6 38	U Gem
U Gem	4 10	U Gem
EX Hya	1 39	U Gem
VW Hyi	1 47	U Gem
RU Peg	8 54–9 03	U Gem
VZ Scl	3 28	U Gem
WZ Sge	1 22	WZ Sge

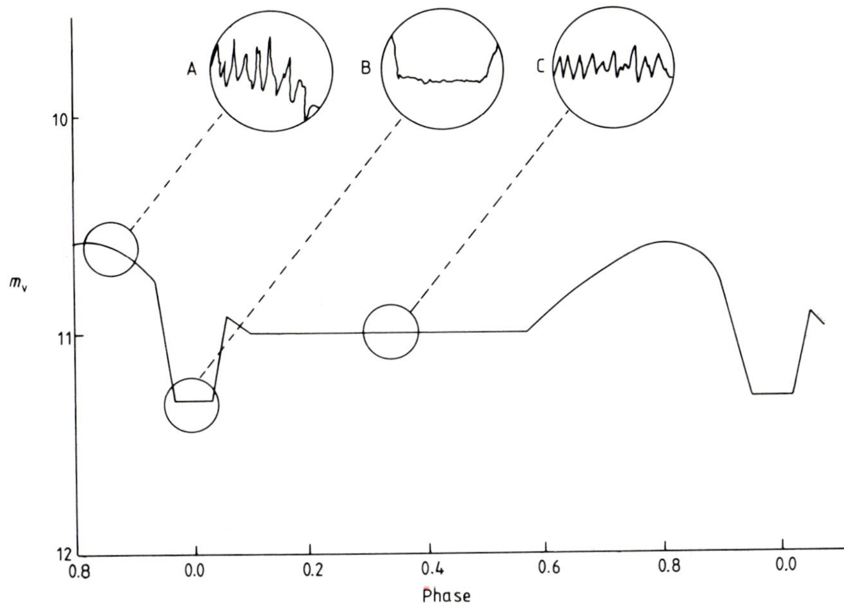

Figure 4.22. Schematic light curve of a dwarf nova throughout its orbital period.

The period of the photometric variations in VW Hyi is 107 minutes at quiescence, and they disappear during a normal outburst. However, in a supermaximum a 110 minute variation is found. Papaloizou and Pringle have proposed two alternative hypotheses to explain this change. In the first, the longer period is the orbital period, and the shorter period the rotational period of the white dwarf (Papaloizou and Pringle 1978a). In the second, the orbital period is 107 minutes and the longer period is due to a modulation of the mass transfer rate caused by a periodic variation in stellar separation, a slight eccentricity in the orbit being the cause of the change in the distance between the stars (Papaloizou and Pringle 1979, Prialnik *et al* 1978). A similar although less pronounced effect is observable in WZ Sge. There is some evidence that the orbital period of U Gem is increasing (Arnold *et al* 1976), but since this requires mass transfer away from the white dwarf it must be regarded with some caution.

At 'normal' minimum (i.e. not an eclipse minimum) the absolute visual magnitude is about $+7.5^{\mathrm{m}}$ (Kraft and Luyten 1965), compared with $+4^{\mathrm{m}}$ for other types of novae. The total amount of energy emitted during an outburst is 10^{31} to 10^{32} J.

Infrared light curves of EX Hya and VW Hyi have been obtained by Sherrington *et al* (1980). That of EX Hya shows the same periodicity as the optical light curve. The 67 minute modulation of the V light curve (Vogt *et al* 1980) for this system is not found in the infrared, and is therefore interpreted as tidal resonance in the hot part of the disc (Papaloizou and Pringle 1980). The binary light curve does not show up in the infrared for VW Hyi. The continuum spectra fit a $\nu^{1/3}$ law which may be expected from an optically thick accretion disc, so that the infrared emission probably originates far out from the binary and is unaffected by its motion. The disc seems likely to fill the white dwarf's Roche lobe (Sherrington *et al* 1980). Dwarf novae in general do not have an infrared excess during their decline after an outburst (Szkody 1977) (in contrast to that normally found for classical and recurrent novae).

BV Cen, EX Hya and VW Hyi have been observed spectrophotometrically in the ultraviolet. BV Cen is faint, but the other two dwarf novae radiate strongly (10^{-15}–10^{-14} W m^{-2} nm^{-1}) with spectral indices of 0.2 ± 0.1 for EX Hya and 0.35 ± 0.05 for VW Hyi (Bath *et al* 1980). SS Cyg has also been detected in the far ultraviolet both when quiescent and during an outburst (Holm and Gallagher 1974).

At soft x-ray wavelengths, SS Cyg, AY Lyr and SU UMa have been detected when they were quiescent (Cordova and Garmire 1979, Cordova and Mason 1980, Fabbiano *et al* 1978), and a weak soft x-ray source may be identifiable with EX Hya (Cordova and Riegler 1979) also during quiescence. A search for x-ray emission during an outburst detected only two out of the twenty dwarf novae covered. These were U Gem and SS Cyg. U Gem reached 2000 counts m^{-2} s^{-1}, and SS Cyg 400 counts m^{-2} s^{-1} for the 0.18 to 0.43 keV region. All the other dwarf novae were undetected at upper limits which ranged from 10 to 160 counts m^{-2} s^{-1}. The x-ray emission from SS Cyg showed the nine second period characteristic of its optical flickering (Cordova *et al* 1980a, Mason *et al* 1978a). Both U Gem and SS Cyg are weak hard x-ray sources (Haefner *et al* 1977, Swank *et al* 1978) during their outbursts. In SS Cyg there is an inverse correlation between the optical and the x-ray intensities. EX Hya is one of the candidates for the optical counterpart of the hard x-ray source 2A 1251 − 290 (Watson *et al* 1978b).

The general form of the light curve is explicable by a model (discussed in more detail later in this chapter) in which the main source of the light is the accretion disc and hot spot formed by the stream of material impinging on to the white dwarf from its companion. The source of the x-ray emission seems likely to be the boundary layer where the accretion disc grazes the surface of the white dwarf (Pringle 1977). The anti-correlation of optical and x-ray emission in SS Cyg is then due to the increase in the optical depth of the accretion disc when the mass transfer rises from 4×10^{13} kg s^{-1} to 2×10^{16} kg s^{-1} during one of its outbursts (Ricketts *et al* 1979).

4.2.2. Spectrum

The main difference between the spectra of novae and dwarf novae is that in the latter the emission lines tend to be slightly stronger and of lower excitation (Greenstein 1960). The emission lines weaken only slightly during an eclipse, so that their emission region can only be partially eclipsed (although in SS Cyg, both optical and ultraviolet emission lines disappear during an outburst). The companion of U Gem has been detected spectroscopically in the red and infrared as an M4.5 dwarf (Stauffer *et al* 1979). The peculiar dwarf nova, AE Aqr whose period is only about one hour, has been observed in the ultraviolet (figure 4.23). The principal low ionisation lines (Mg II and Ca II) arise in the outer optically thin parts of the disc, while the high excitation lines (He II, N V, C IV and Si IV) seem to be formed near the centre of the disc. The Lyα and He II 164 lines arise from recombination, the other lines are all collisionally excited (Jameson *et al* 1980).

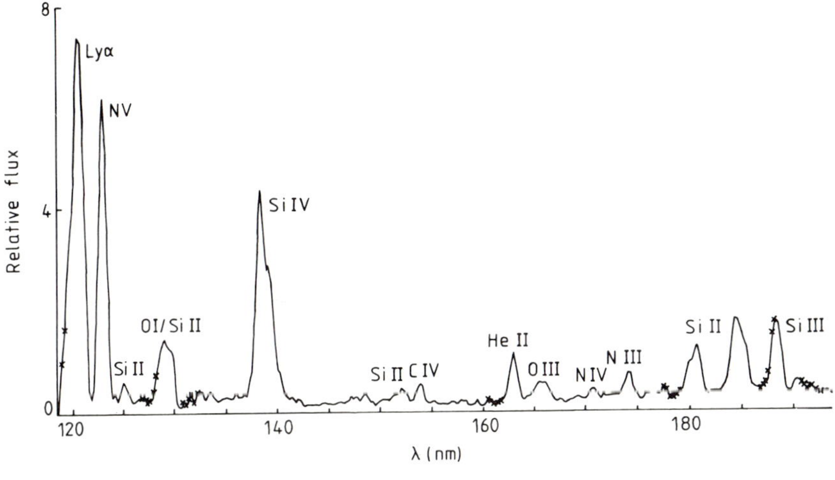

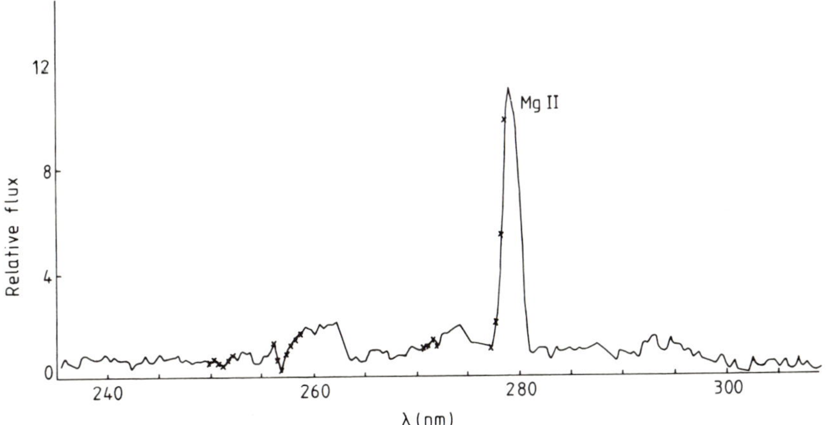

Figure 4.23. Ultraviolet spectrum of AE Aqr. (Reproduced from Jameson *et al* 1980 by permission.)

4.2.3. Masses

The masses of dwarf novae may be obtained in a similar way to those of other novae (see earlier in this chapter), and are listed in table 4.5. EM Cyg is a double-line eclipsing binary whose masses are thereby given as (Robinson 1974)

$$M_2 = 0.90 \pm 0.17\, M_\odot,$$

$$M_1 = 0.70 \pm 0.18\, M_\odot.$$

Table 4.5. Masses of dwarf novae (Bailey and Ward 1981, Breysacher and Vogt 1980, Kiplinger 1979, Ritter 1976, 1980a,b, Robinson 1976b, Robinson *et al* 1978, Warner 1975).

Name	$M_2/M_\odot$	$M_1/M_\odot$	Type
RX And	0.65	1.02	Z Cam
SS Aur	0.57	0.89	U Gem
Z Cam	0.86–0.90	1.17–1.27	Z Cam
OY Car	0.12–0.14	0.30–0.4	
WW Cet	0.50		Z Cam
Z Cha	0.16–0.17	0.35	U Gem
EM Cyg	0.86–0.95	0.67–0.73	Z Cam
SS Cyg	0.83–0.89	0.80–1.07	U Gem
U Gem	0.53–0.55	0.35-0.92	U Gem
EX Hya	0.16–0.19	1.07–1.4	U Gem
VW Hyi	0.17		U Gem
RU Peg	1.08–1.20	0.92–1.02	U Gem
WZ Sge	0.019–0.12	0.38–1.5	WZ Sge

The inclination of Z Cam may be estimated to lie between 52° and 68° from the presence of a hump in its light curve, but the absence of an eclipse. This leads to estimates of the masses in this system of (Warner 1975)

$$M_2 = 0.69 \pm 0.12\, M_\odot,$$

$$M_1 = 0.97 \pm 0.20\, M_\odot.$$

These masses are in good agreement with those obtained by indirect methods and listed in table 4.5, providing support for the validity of those estimates for other systems. There do not appear to be any significant differences in the masses of either of their components between dwarf and other types of novae.

4.3. The Disc

In all types of novae the material accreted on to the white dwarf initially forms a disc around the star (for further discussion of the disc and models of novae see later in this chapter). Some of the properties of the disc may be measured directly. The profiles of the hydrogen emission lines (at minimum) give radii for the disc of 0.4 to 0.66 of the separation of the centres of mass of the components (Robinson 1976a). These figures are over-estimates, however, since Keplerian motion of the gas around the white dwarf is assumed, and this will not be true for material which fills a substantial fraction of a Roche lobe.

The density is highest near the white dwarf, but the bulk of the hydrogen emission comes from the lower density, outer parts of the disc. Estimates of the electron density range from 10^{18} to $10^{27}\,\mathrm{m}^{-3}$, and of the temperature from 7000 to 400 000 K (the latter figure only for the hot spot formed by the impinging stream of material from the companion). The total mass of the disc may be from 10^{-5} to $10^{-4} M_\odot$ (Bath 1977b). The rate of mass transfer has been estimated as 2×10^{-9} to $3 \times 10^{-7} M_\odot\ a^{-1}$ from variations in the orbital periods (Robinson 1976a), but this cannot be regarded as reliable since the change in period is erratic, and there may be both increases and decreases in the period of a single system. Since mass transfer cannot explain both types of change (on a short timescale) other processes must be operating. Rates of less than 3×10^{-8} and $1.3 \times 10^{-9} M_\odot\ a^{-1}$ have been found spectrophotometrically for VW Hyi and EX Hya (Bath *et al* 1980), and a mass transfer rate of $5 \times 10^{-9} M_\odot\ a^{-1}$ for Z Cam (Kiplinger 1980).

4.4. Models for Novae

A model for all novae which fits the binary light curves and provides a mechanism for the outbursts (without, however, reproducing all the features of the light curves of outbursts) is now fairly generally agreed (figure 4.24).

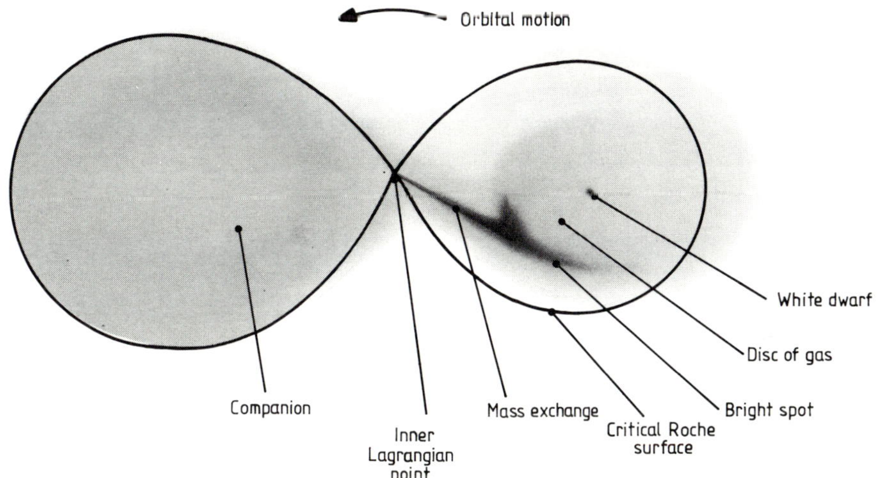

Figure 4.24. Schematic view of a nova system.

The system is a close binary with the original primary evolved to the white dwarf stage, and the original secondary (referred to as the companion throughout this chapter) evolving away from the main sequence. For the very closest systems the companion may still be a main sequence star. The companion fills its Roche lobe, and material is spilling from it down towards the white dwarf. The transferred material congregates in a disc around the white dwarf and partially fills its Roche lobe. A fraction of the material may spill out of the Roche lobe and be lost to the system as a continuous mass loss. The interaction zone between the infalling material and the disc forms a hot spot (Flannery 1975). At 'normal' minimum (i.e.

not an eclipse minimum) the light from the system is almost entirely due to the hot spot. Neither the white dwarf nor its companion contribute more than a small fraction to the light of the system. Recurrent novae and one or two peculiar novae with giants for companions are an exception to this, as mentioned earlier. A possible cause of the difference between classical and dwarf novae may lie in the degree of evolution of the companion. In classical novae it is less evolved and more massive than in dwarf novae, resulting in mass transfer rates up to three times higher in a classical nova than in a dwarf nova of the same orbital period (Whyte and Eggleton 1980).

The binary light curve (figure 4.22) is hence due to the motion of the hot spot. The hump occurs as the hot spot is viewed directly, and if the inclination is sufficient the eclipse occurs as the hot spot passes behind the companion. Outside phases 0.6 to 0.1, the hot spot is viewed from 'behind', and is partially obscured by the disc of material. The light curves of classical novae do not in general show humps or eclipses, but a hump is just detectable in the light curve of T Aur. In this case the total brightness of the disc may be greater than that of the hot spot (Bianchini 1980).

The flickering originates in the hot spot, as may be inferred from its disappearance during eclipse, and its greater amplitude during the hump. The 360° phase shift when it is visible during eclipse must reflect a variation in the phase of the flickering across the hot spot. The reason for the flickering is not completely clear. Its stability in DQ Her argues strongly for a mechanism based on the rotation of the white dwarf. Katz (1975) has produced a model with one bright magnetic pole and a 71 second revolution for the white dwarf, which receives support from its additional ability to explain the asymmetric wavelength dependence of the pulsed amplitude of the He II 468.6 line (Ali-Alper 1979). Petterson (1980) suggests a progradely rotating beam from the white dwarf which is reflected by a large accretion disc. A magnetic field of 10^2–10^3 T is indicated by the polarisation of this system (Rose and Scott 1976), which could interact with the disc to produce the flickering by non-radial pulsations of the disc (van Horn *et al* 1980). In other systems the flickering may have several periods and/or change period and/or last only a short while. In such cases a different mechanism seems to be required, and non-radial oscillations of the white dwarf (Nather and Robinson 1974, Papaloizou and Pringle 1978b, Warner and Robinson 1972), or the eclipse by it of transient orbiting bright condensations (Bath 1973) or vortices (Haefner *et al* 1979) in the disc have been suggested. Sparks and Kutter (1980) postulate an accretion-driven wave moving around the white dwarf which is alternately driven and damped to explain the period changes.

The outbursts seem likely to arise from thermonuclear runaway reactions at the surface of the white dwarf (or perhaps in the accretion disc). The accreted matter is largely hydrogen and will accumulate on the surface of the white dwarf at high densities (10^7 kg m^{-3} or more). When perhaps $10^{-4}\,M_\odot$ has accumulated, the density and temperature will be high enough for the nuclear burning of hydrogen. The temperature will then rise rapidly without a corresponding pressure rise (because the material is degenerate), and so the hydrogen shell will not be disrupted initially. The rates of nuclear reactions are very sensitive to temperature, and so will increase extremely rapidly, raising the temperature to 10^8 K or more. The degeneracy will be lifted at such a temperature, leading to the final explosion which completely disrupts the hydrogen shell around the white dwarf (figure 4.25). The over-abundance of carbon, nitrogen and oxygen in the ejecta is strong evidence in favour of this process (Starrfield *et al* 1979). In fast classical novae the material must be enhanced by factors of 20 to 100 in these elements in order to reproduce the main features of the light curve. In slow novae the outburst can occur in unenhanced material only if substantial contributions to the energy are made by dynamical friction between the expanding envelope and the companion. Only in the very slow peculiar novae, RR Tel and RT Ser, can the outburst occur in unenhanced material without further energy sources (MacDonald 1980). In recurrent novae, enrichment of the material from the red giant in ^{3}He may aid the outburst (Shara 1980). A schematic indication of the relationship between envelope mass, enrichment and nova type is shown in figure 4.26. (An extreme ultraviolet (EUV) object does not have mass ejection, and appears as a very hot object radiating mostly in the extreme ultraviolet (Shara *et al* 1980).) The β-unstable nuclei, ^{13}N, ^{14}O, ^{15}O and ^{17}F are convected into the reaction region where their decay adds to the energy release (Starrfield 1980) (which may reach 10^8 to 10^9 W kg^{-1} at maximum). Later in the outburst they provide a continuing source of energy.

The material is driven outwards by shock waves, gas pressure and radiation pressure. Different mechanisms may drive the material at different phases of the outburst (and may vary between types of nova). In the model of fast novae

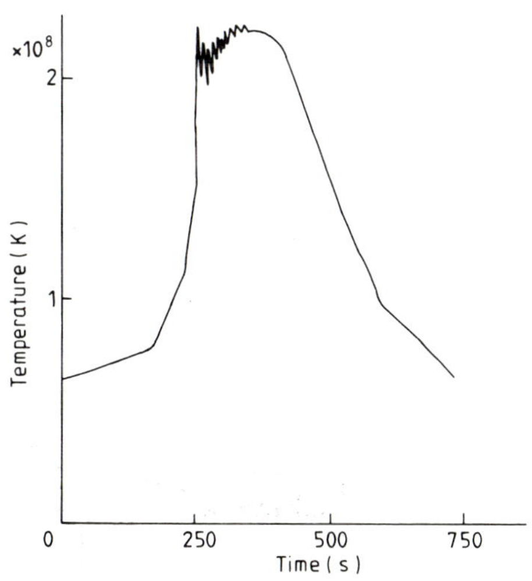

Figure 4.25. Temperature of the hydrogen shell (with 20% carbon and oxygen) during a nova outburst. (Reproduced from Starrfield *et al* 1976 by permission.)

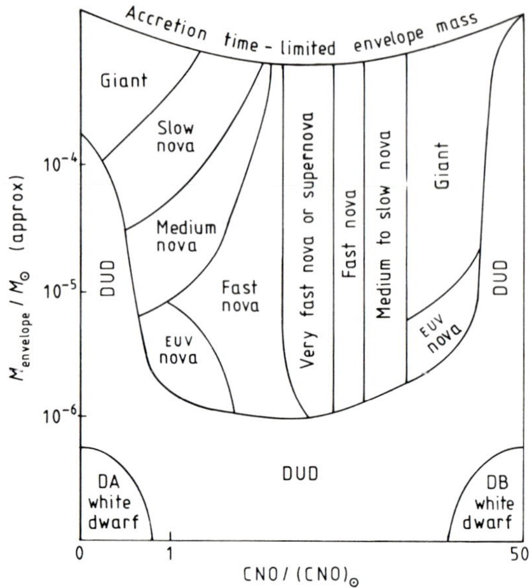

Figure 4.26. Semi-quantitative relationship between the envelope mass and its enrichment in carbon, nitrogen and oxygen, and the type of nova. (Reproduced from Shara *et al* 1980 by permission.)

proposed by Prialnik *et al* (1979) the nuclear runaway occurs away from the core–envelope interface, and a burning front recedes through the envelope. The nuclear reactions are halted, perhaps after about 30 days, by the decrease in density due to the rapid expansion. About half the envelope material is completely ejected. Their slow nova model (Prialnik *et al* 1978) has a lower mass star ($0.8\,M_{\odot}$), and the outburst declines due to exhaustion of material. Up to 95% of the envelope may be ejected. The initial rise in brightness corresponds to the physical expansion of the region where $\tau = 1$. The decline in the visual brightness seems then to be due to the redistribution of the energy to shorter wavelengths (Gallagher and Starrfield 1976). The transition period occurs when the expanding material becomes optically thin, so that the inner, unstable, parts of the outburst (where ejection is still occurring) are visible. Alternative hypotheses are that the transition phase arises from dust formation or from a hydrostatic collapse after a reduction in radiation pressure (Phillips and Selby 1977). The constant (total) luminosity phase is due to nuclear burning at the base of the envelope, of unejected envelope material. The nova returns to quiescence when these reactions cease, perhaps one to ten years after the outburst (Starrfield 1980). The light curve for the model whose temperature curve was shown in figure 4.25, is shown in figure 4.27. It is possible that slow novae occur on low-mass white dwarfs, while fast novae occur on high-mass white dwarfs (Shara 1980, Starrfield *et al* 1976). The possibility of a quasi-steady optically thick stellar wind occurring after the initial eruption has been suggested by Bath and others (Bath 1978, Bath and Shaviv 1976, Hartwick and Hutchings 1978). This

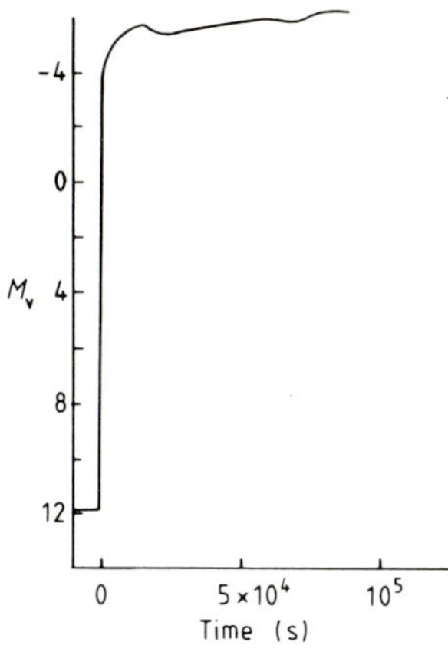

Figure 4.27. Theoretical light curve of a nova (for the same model which produced the temperature curve shown in figure 4.25). (Reproduced from Starrfield *et al* 1976 by permission.)

models the decline of a classical nova by allowing the mass loss rate to decrease slowly, and the wind to become progressively optically thinner.

The near disappearance of the eclipses in U Gem during an outburst, argues that the bright spot is essentially unchanged, and that it is the disc of gas from which the energy is being emitted. In Z Cha the white dwarf is eclipsed and, as noted earlier, the light curve throughout an eclipse becomes more rounded during an outburst than when the system is quiescent. Such behaviour suggests that the eclipse is partial and the size of the stars or eclipsing bodies is changing (just such a change has been observed by Bateson (1978)). In one star, VW Hyi, the hot spot does increase in brightness during an outburst (Haefner *et al* 1979). This may be due to an increased rate of mass transfer from the companion, stimulated by the outburst itself.

The expanding nebula may sometimes be photographed directly, some years after the outburst. The nebula is rarely circular (or spherical), more usually it is bipolar or irregular. Such shapes may be attributable to interaction of the expanding material with the disc or the companion (Hutchings 1974).

Thermonuclear runaway in the outer layers of the white dwarf seems adequate to explain most of the features of novae. However, it has drawbacks; in particular the timescale for the decline should be comparable with the cooling timescale of a white dwarf, which is far longer than the observed time for decline. Theories of the evolution of close binaries (Paczynski 1971) suggest that the white dwarf will be a helium white dwarf. The thermonuclear runaway, and the over-abundance of carbon, nitrogen and oxygen require it to be a carbon–oxygen white dwarf (Colvin *et al* 1977, Starrfield *et al* 1974). Other mechanisms have therefore been suggested, and these may be important in dwarf novae. The outburst may be caused by a sudden increase in the mass exchange rate which results in the whole disc brightening. Since 10^{13} J are released when 1 kg of material falls on to a white dwarf, this mechanism could provide sufficient energy. It is difficult, however, to see how it could produce the extreme rapidity of the outbursts. Nonetheless it remains a possibility to explain the outbursts in U Gem stars where there may be no net mass loss to the system. During a supermaximum of a dwarf nova the mass flux is about four times higher than in a normal maximum (Mayo *et al* 1980), which fits in well with this suggestion. If the material is not ejected then classical nova outbursts may be expected in dwarf novae at similar intervals (10^5 years) to those in other novae. A brightening of the accretion disc may also explain the 1978 outburst of WZ Sge (Crampton *et al* 1979). Alternatively a sudden increase in the mass exchange rate may strip off the outer layers of the companion, so that the nova is due to a brightening of the companion and not of the white dwarf or disc. The peculiar slow nova, RR Tel, may be a candidate for this mechanism. A further, remote, possibility which could result in novae on single white dwarfs, is the accumulation of interstellar material on the surface of the white dwarf, followed by thermonuclear runaway. The white dwarf would need to pass through a very extensive, and dense interstellar gas or dust cloud for such a situation to arise, and no novae are currently attributed to this mechanism. Collapse of a magnetic white dwarf, leading to the expulsion of material at the magnetic poles is another mechanism that has been suggested (Collins and Foltz 1977).

4.5. Symbiotic Stars

Symbiotic stars have a number of behavioural characteristics which are somewhat nova like, so that it is appropriate to mention them in this chapter. They have combination spectra, which at minimum brightness look like that of a late type giant. As the star brightens this becomes overlaid by a high-excitation emission line shell spectrum. Type I symbiotics (e.g. Z And) have quasi-periodic changes in brightness with a timescale of a few months. At minimum their spectra combine those of a cool giant with a high excitation spectrum. Type II (e.g. V 1016 Cyg) look like long-period variables before outburst, with low-excitation emission lines (Paczynski and Rudak 1980). They are usually interpreted as semidetached binary systems, with a long-period variable filling its Roche lobe and with a hot blue star as the other component. The outbursts have an amplitude of about six magnitudes, and a steeper increase in brightness than the decline (figure 4.28). At least one, CH Cyg, has shown nova-like flickering during an outburst (Slovak and Africano 1978), but others do not (Walker 1977). It is possible that mass exchange occurs as a stellar wind, or by isothermal pulsations, so that there is no disc or hot spot around the hot component. The mass exchange is usually thought to be towards the hot component, although Hutchings *et al* (1975) have interpreted AG Peg on the basis of mass transfer to the cool component. The outbursts may arise from variations in the accretion rate due to intrinsic instabilities in the disc, or mass transfer instabilities in the red giant (Bath 1977a).

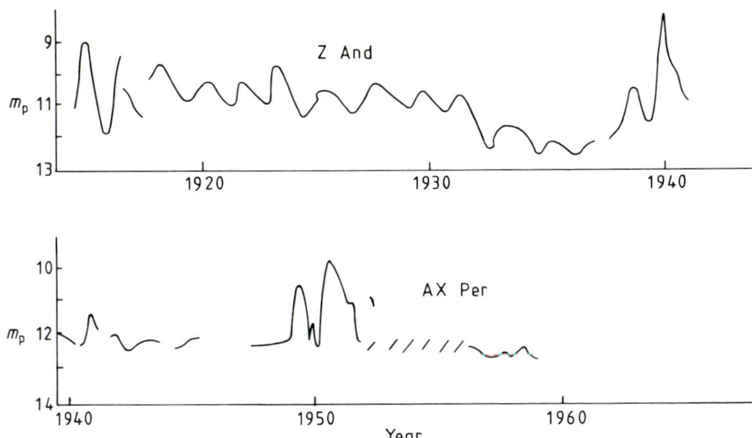

Figure 4.28. Light curves of symbiotic stars. (Reproduced from Boyarchuk 1969 by permission.)

Free-free radio emission at 15 GHz has been detected from nine symbiotic stars. Most of their spectral indices are near +0.6 which is indicative of mass loss on a scale comparable with the Mira variables (Wright and Allen 1978). AG Peg emits free-free radiation with an intensity of about 10 mJy from 8 to 10 GHz, leading to an estimate of its mass loss rate of $10^{-6} M_{\odot}\ a^{-1}$, and a mass for its nebula of $7 \times 10^{-5} M_{\odot}$ (Gregory *et al* 1977).

There is an overlap with other types of object – T CrB and RS Oph are recurrent novae and type II symbiotics. 17 Lep is also classified as a Be star as well as a symbiotic variable. They are among the systems suggested for planetary nebula precursors (chapter 3).

4.6. Transient X-ray Sources

X-ray emission has been found from novae and dwarf novae, and some aspects of its behaviour have been discussed earlier. There are a number of more tentative correlations for which the identification of the x-ray source with the nova is not unambiguous. Mass exchange in close binary systems is a common model for x-ray sources, and the behaviour of the transient sources; a sudden brightening and a rapid decay, with 10 to 20 days at maximum, is very nova-like. The identifications are therefore probably more certain than might be indicated on the basis of positional coincidences alone.

H 2252–033 has been reasonably positively identified with a 13th magnitude emission line object which appears similar to a dwarf nova at minimum (Griffiths *et al* 1980). The optical counterpart of Sco X 1 shows nova-like flickering when it is active, and this is correlated with its x-ray flickering (Ilovaisky *et al* 1980). Nova Mon 1975 and the transient x-ray source A 1524–61 have been mentioned earlier. They do not appear to be completely normal novae, and other suggestions for their interpretation include a neutron star variant of a dwarf nova, an extreme classical nova, or an extreme transient x-ray source (Carpenter *et al* 1976, Lloyd *et al* 1977, Long and Kestenbaum 1978, Murdin *et al* 1977, Whelan *et al* 1977). Other possible candidates for the x-ray counterparts of novae include Cen X 2, Cen X 4, Cet X 2, Cep X 4, 2U 1543 − 47 and A 1118 − 61 (Warner 1975). However, one Sco X 1 type source, 4U 1735 − 44, exhibits type 1 bursts (White *et al* 1980a), and these are generally attributed to accretion on to a neutron star, and not on to a white dwarf (Swank *et al* 1977). A number of cataclysmic variables are also found in the error boxes for the sources of γ-ray outbursts (Vidal and Wickramasinghe 1974).

The recently recognised AM Her group of stars (also known as 'polars') also show affinities with novae. They are short-period binaries (Priedhorsky *et al* 1978a) whose optical and infrared light curves show the flickering characteristic of novae (Jameson *et al* 1978) (figure 4.29). In Am Her, the period of the flickering is about 80 seconds (Szkody and Margon 1980). Their brightness also varies more irregularly by up to two magnitudes with timescales of hundreds of days. Their spectra contain strong variable emission lines due to H, He I, He II and C III (Schneider and Young 1980) (figure 4.30). The He I 447.1 and He II 468.6 emission lines in AM Her have two components; broad and sharp. The velocity amplitude of the former is four times that of the latter, and is out of phase with it and with the eclipses by 133° (Greenstein *et al* 1978). In one system (VV Pup) there are very broad absorption features which may be high harmonics of cyclotron absorption (Visvanathan and Wickramasinghe 1979). They are also weak x-ray sources (Chiappetti *et al* 1980). In fact the first interest in AM Her arose through its identification as the optical counterpart of 3U 1809 + 50, and most other stars in this class have been similarly identified from x-ray sources. AM Her is an eclipsing system in the x-ray region as well as in the optical. One of the more unusual features of the class is the behaviour

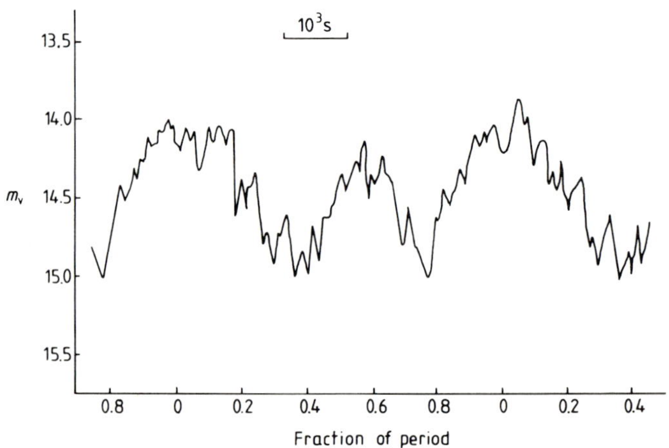

Figure 4.29. Light curve of the AM Her binary 2A 0311 − 227. (Reproduced from Watson *et al* 1980 by permission.)

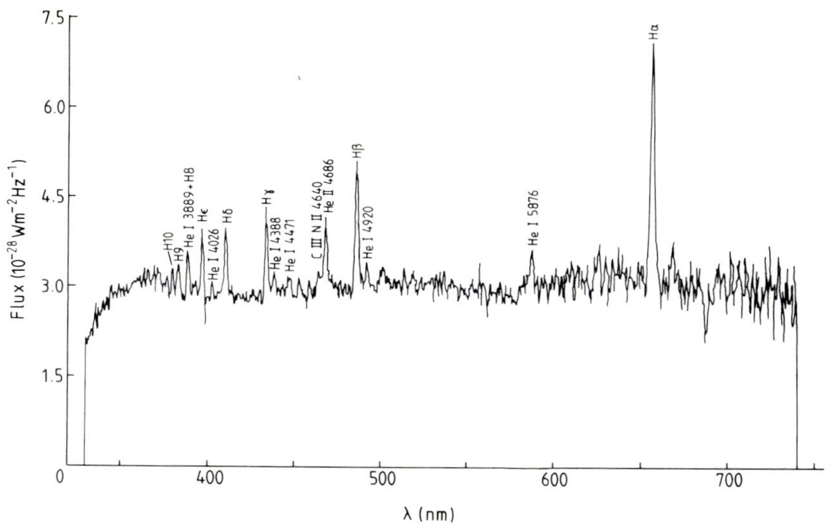

Figure 4.30. Spectrum of the AM Her type optical counterpart of 2A 0526 – 328 (also classified as an UX UMa type system (Warner 1980)). (Reproduced from Watts *et al* 1980 by permission.)

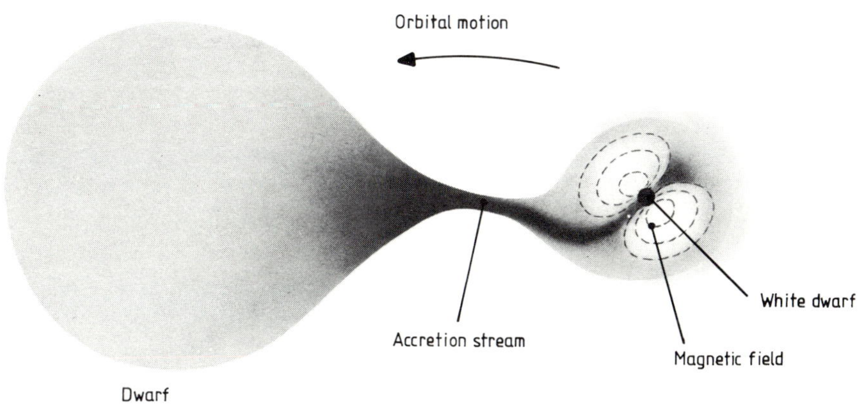

Figure 4.31. Schematic model of an AM Her type system.

of their polarisation. Normally this is barely detectable, but it rises sharply and briefly to 2 to 20% at the time of the x-ray eclipse (Bailey *et al* 1978, Tapia 1977). Even more unusual is that variable circular polarisation is found whose period is also that of the binary, and which flickers in phase with the optical flickering (Stockman and Sargent 1979). In VV Pup the linear polarisation peaks at 7 to 9% when the system is faint, but reaches 16% when it is bright (Bailey 1978). The circular polarisation also varies, being negative when the system is bright, but otherwise almost absent (Liebert and Stockman 1979). The unusual polarisation behaviour is the reason for the alternative name of polars for this group of stars.

The model for the group is again a close binary with a cool Roche lobe-filling dwarf with a white dwarf of about one solar mass (Chanmugen and Wagner 1977, King and Lasota 1979, Liebert *et al* 1978, Priedhorsky *et al* 1978b). Material is lost from the dwarf to the white dwarf, but does not form an accretion disc. Instead a very strong magnetic field (10^3 to 10^4 T) channels the flow of material towards the magnetic poles. Most of the radiation from the system, at all wavelengths, is then produced in the accretion streams (figure 4.31).

5. *Be Stars*

5.1. Introduction

The earliest recorded observation of an emission line in a stellar spectrum was by Secchi (1866) who found Hβ in emission in the Be star γ Cas. Be stars, and in particular γ Cas, have continued to interest astronomers ever since. Thirty years ago their basic model seemed clear, and only the details remained to be completed. However, the last decade has put everything back into the melting pot, and there are now several competing models, none of which can completely fit all the recently acquired data.

The strict interpretation of the spectral class Be, is just a B type star with emission lines. However, tradition has reserved this term for a subset of such stars, and also includes in it the few stars of similar characteristics but of spectral classes O and A (occasionally called Oe and Ae stars). The Be stars comprise about 10 to 20% of the B type stars with a maximum proportion at about spectral class B3. The main group of stars excluded from consideration by this definition are the supergiants (which are discussed briefly in chapter 6). The supergiants number about 10% of the B emission stars, and are separable from other types of Be star by their Balmer decrement (Crawford *et al* 1975). Another subset, the Herbig Ae and Be stars, is considered separately in the final section of this chapter. The B[e] stars are defined (Conti 1976) as B type objects which show forbidden optical emission lines, and they are an extremely disparate group. Their absolute visual magnitudes range from -1.5 to -10 (Allen 1979a). η Car, the brightest of them is discussed further in chapter 6.

Confusion has long existed between Be stars and shell stars. As recently as 1976, Hutchings (1976b) was able to introduce a review at an IAU symposium on Be stars and shell stars with:

> 'The distinction between Be stars and shell stars has often been a confusing one to me, and I can find no clearly defined criteria in the literature. I suspect that the confusion is not restricted only to myself. . . .'

It is time that this muddled classification was dropped, and shell stars, shell phases, and shell spectra etc will be considered to be aspects of Be star behaviour in this work. An attempt to provide a unified classification has recently been made by Jaschek *et al* (1980). They identify five sub-groups of Be stars whose identifying characteristics are:

Group I: Fe II in emission
Balmer emission extending to high members of the series
Possible He I 587.6 emission
Spectral type B0 to B6.

Group II: Hα, Hβ emission with sharp absorption cores which are permanent or semi-permanent
Spectral type B3 to B8.

Group III: Only Hα and Hβ have emission components
Sharp absorption cores to the higher Balmer absorption lines
Narrow variable metallic absorption lines
Spectral type B5 to A0.

Group IV: Hα, Hβ permanently in emission (plus Hγ for the earliest spectral types)
Hβ always has emission superimposed upon a broad absorption line
Spectral types B3 to A0.

Group V: B → Be star variables (or vice versa). They may belong to any of the above groups during their Be phase(s).

This seems likely to be a useful system, but has not been in existence long enough to have proved its worth.

5.2. Spectral Features

The main characteristics used to define a Be star are:

(*a*) There is an absorption and continuum spectrum that resembles that of a normal B type star, except that the absorption lines are often broad and diffuse. The widths of the absorption lines are often wavelength dependent.

(*b*) There is an emission line, or lines, whose width(s) is smaller than that of the broad absorption lines. The emission lines are frequently superimposed on the broad absorption lines, usually symmetrically, but occasionally displaced to the violet or red. The emission lines may be single or double (or very occasionally more complex), and examples of profiles are shown in figures 5.1 and 5.2.

Other significant features found to occur more or less regularly in Be stars are:

(*a*) Abnormally narrow absorption lines. These may be absorption cores to emission or broad absorption lines, or they may be simple absorption lines. They tend to be present as simple absorption lines mainly for the metals and in the ultraviolet, while they are usually absorption cores for the hydrogen lines. Some authors use the presence of such lines to indicate the presence of a shell star or a shell phase.

(*b*) *V*/*R* variation. The quantity *V*/*R* is the ratio of the intensity of the violet component of a double emission line to the intensity of its red component. It may vary between zero and infinity (i.e. the violet component may vanish completely

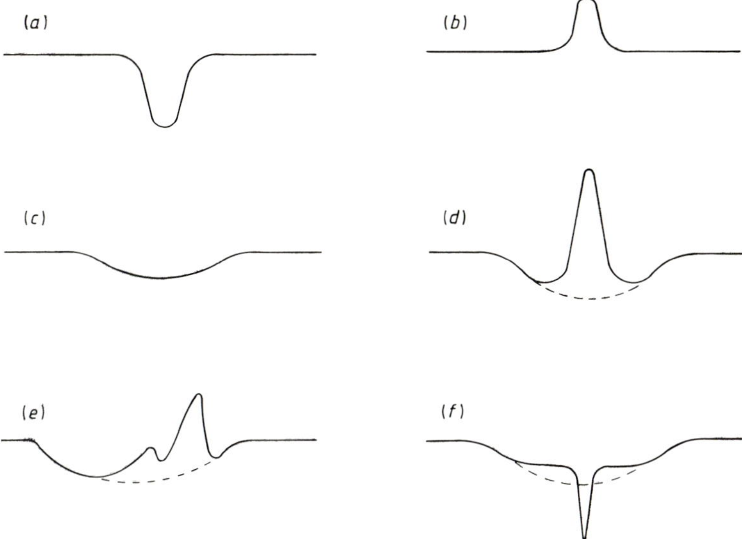

Figure 5.1. Schematic line profiles: (*a*) absorption line in a Be star, (*b*) single emission line in a Be star, (*c*) broad absorption line in a Be star, (*d*) superimposed emission and absorption lines in a Be star, (*e*) asymmetrically superimposed absorption and double emission lines in a Be star, (*f*) superimposed broad absorption, incipient emission and central absorption lines in a Be star.

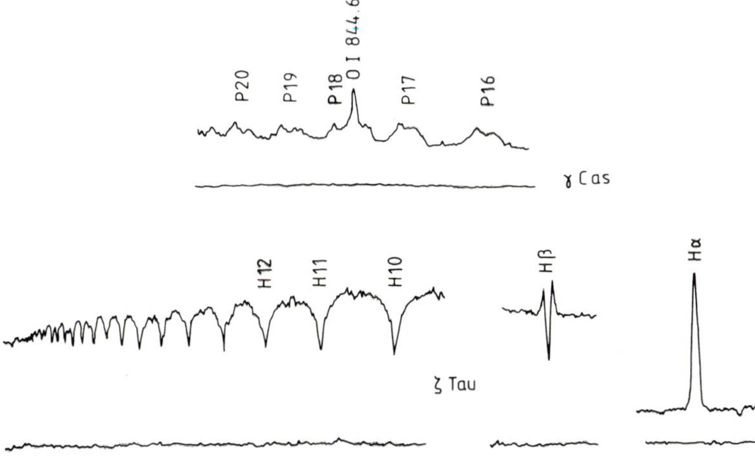

Figure 5.2. Be star spectra.

leaving only the red, or vice versa). Variations on a smaller scale than this are extremely common in Be stars. The timescale of the variations ranges from minutes to years, with the larger amplitude variations usually taking at least days or weeks to occur. Occasionally the variations may be quasi-periodic, or they may start up in a previously stable spectrum, or stop after occurring for many years (figure 5.3).

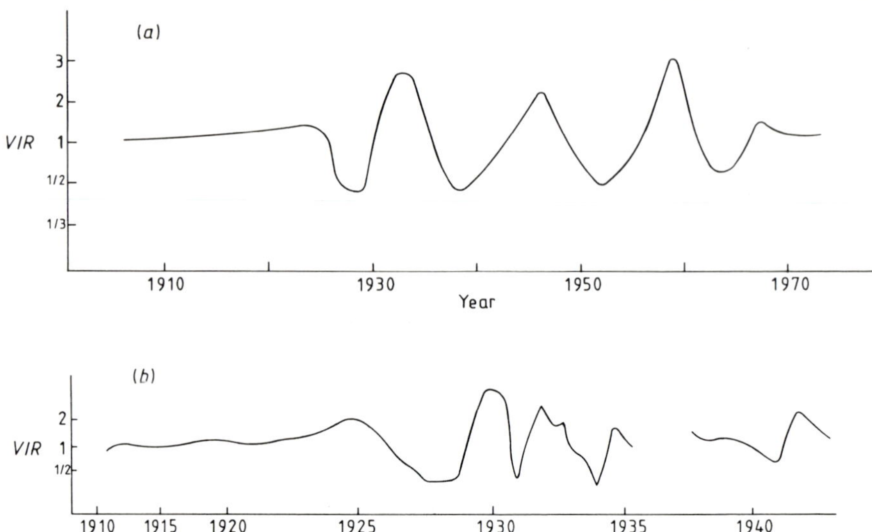

Figure 5.3. *V/R* variations of Hβ in Be stars (*a*) β' Mon, (*b*) π Aqr. (Reproduced from Hutchings 1976b, McLaughlin 1961 by permission.)

(*c*) *E/C* variation. The quantity *E/C* exists for all emission lines and is the ratio of the intensity of the emission line (or the sum of the intensities of the emission components) to the intensity of the nearby continuum. It is similar to the equivalent width of the line, except that it is usually an eye estimate, not a measured and calculated quantity. The *V/R* and *E/C* variations often follow similar patterns, but this is not invariably the case. Some typical variations are shown in figure 5.4. Minor variations can occur within a few minutes (Bahng 1976, Doazan 1976, Hutchings 1969).

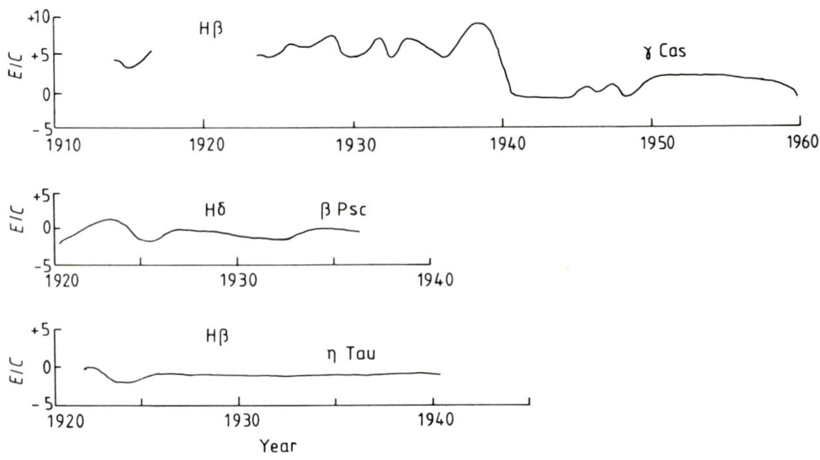

Figure 5.4. *E/C* variations in Be stars.

(*d*) Absorption line variations. The two types of absorption line, broad and narrow, which were discussed above behave quite differently. The broad lines change very little, while the narrow lines may vary very considerably. Their variations often correlate with the E/C changes, the narrow absorption lines strengthening or weakening as E/C increases or decreases.

(*e*) Velocity variations are frequently present. Many Be stars are binaries (e.g. ϕ Per and ζ Tau, see figure 5.5), and their velocity curves are strictly periodic. Others, e.g. 25 Ori (Dodson 1936), and 48 Lib (Faraggiana 1969), show only semi-regular variations with timescales from a few months to several years. The velocity changes, like the other changes (except when they arise from binary motion) may stop and start. For example, the semi-regular velocity variation in HD 20336 which ceased completely in 1938 (figure 5.6).

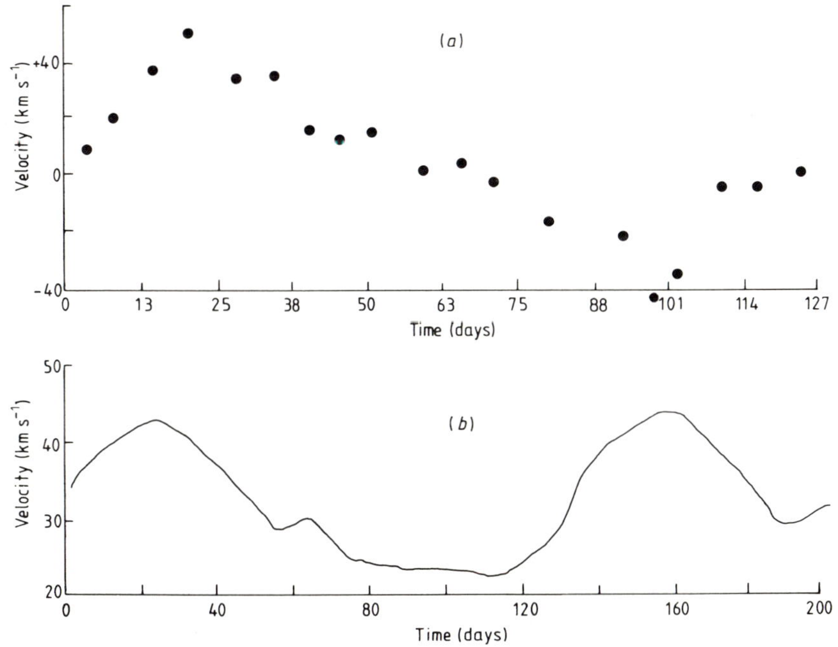

Figure 5.5. Velocity variations in ϕ Per and ζ Tau. (Reproduced from Hendry 1976, Losh 1931 by permission.)

(*f*) About 70% of Be stars show small (0.1^{m}), possibly periodic, magnitude variations. Larger changes are rarer, but possible, γ Cas for example changed by over one magnitude between 1937 and 1940.

A sub-class of these stars, called Bex (for extreme Be star) has been identified by Schild; these have stronger emission and weaker absorption lines than average, and are slightly redder than an ordinary Be star.

In spite of the apparent diversity of the features in the spectrum of a Be star, they form quite a well defined group of stars. Since they comprise 10 to 20% of the

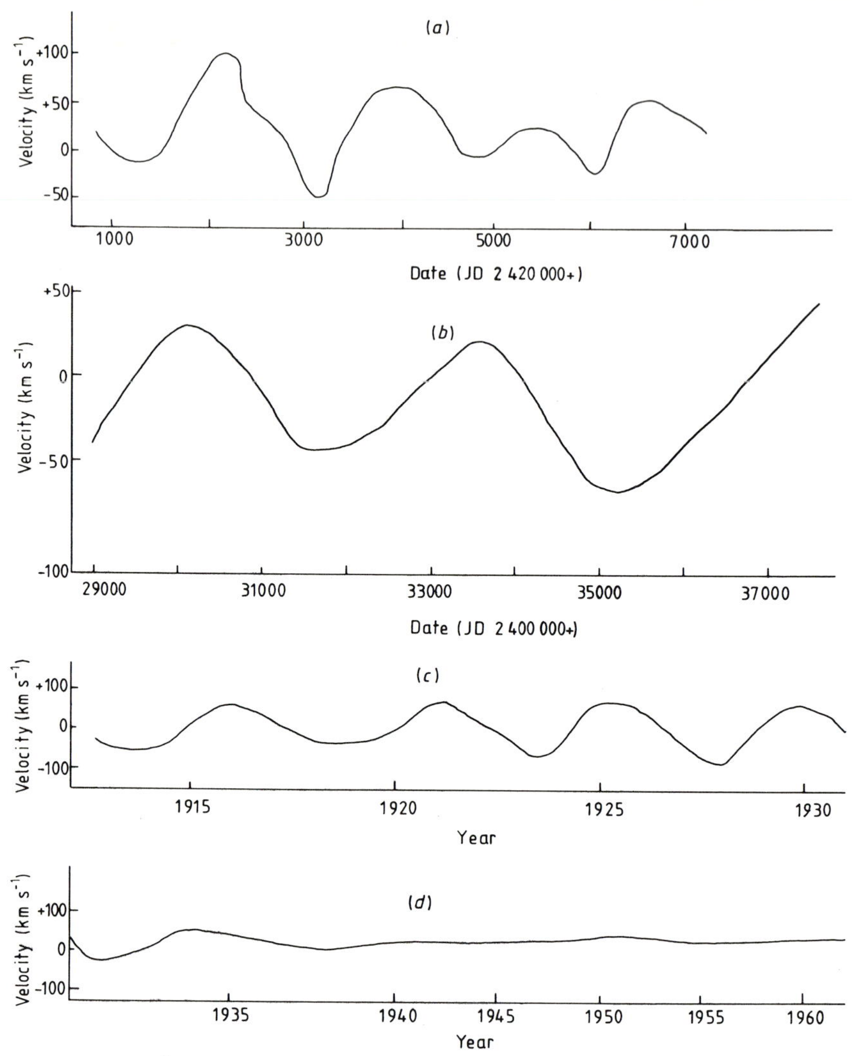

Figure 5.6. Velocity variations in 25 Ori, 48 Lib and HD 20336. (Reproduced from Dodson 1936, Faraggiana 1969, McLaughlin 1963 by permission.)

B type stars, very large numbers of them are known. Numerous catalogues exist, of which the most useful is the Mount Wilson catalogue (MWC) (Merrill and Burwell 1933, 1943, 1949, 1950). Bidelman (1976) lists shell stars, Bertiau and McCarthy (1969) and Jaschek *et al* (1971) list Be stars with bibliographies. Catalogues of emission line objects (which include many types of star, not just Be stars) include those of Wackerling (1970), Stephenson and Sanduleak (1977) and Henize (1976).

5.3. Rotation

The suggestion that rotation might be determined from spectral line profiles was made early in the history of spectroscopy, Captain W de W Abney (1877) suggesting that:

> 'The advancing limb . . . will cause the absorption lines to move towards the violet end, and the receding limb towards the red end of the spectrum.'

In Be stars the width of the broad absorption lines is often proportional to their wavelength (this may also be true for the emission lines on occasion). Struve (1931) found for the emission lines that their width, $\Delta\lambda$, was given by:

$$\Delta\lambda = 6.28 \times 10^{-3}(\lambda - 327.0)(W - 0.261) + 0.261 \quad (5.1)$$

where λ is the wavelength of the line and W is the measured width of the Hβ emission. This dependence of line width on wavelength strongly suggests Doppler broadening as the principal broadening mechanism. Struve (1951) also showed that the Doppler effect producing this broadening was not that of turbulence or thermal motions of the atoms, thus leaving rotation as the only likely source for Doppler broadening.

A method for determining $v \sin i$ was proposed by Shajn and Struve (1929). (The quantity, $v \sin i$, is the apparent equatorial rotational velocity of the star, v is the actual equatorial rotational velocity, and i is the angle of inclination of the star's rotational axis to the line of sight.) This method was corrected for the effects of limb darkening, gravity darkening, non-spherical stars etc by Stoeckley (1968) and others. Other approaches to determining the rotational velocities of Be stars have been line profile synthesis from model atmospheres (Collins 1974), Fourier transformation (Gray 1973), and attribution of the cause of periodic spectral and photometric changes to rotation (Hutchings 1970b, Walker 1953). The resulting values of $v \sin i$ have been tabulated by Slettebak (1976), and their distribution is shown in figure 5.7.

A random distribution of stellar rotational axes for a set of stars with a given equatorial rotational velocity, V_0, produces a sharply peaked distribution curve. Consider a star whose angle of inclination to the line of sight is i. Then the observed rotational velocity, V, is:

$$V = V_0 \sin i \quad (5.2)$$

and the probability that V will lie in the range:

$$v + dv \geqslant V \geqslant v - dv \quad (5.3)$$

will be proportional to the area of the shaded fraction of the sphere in figure 5.8, which is defined by:

$$\alpha + d\alpha \geqslant i \geqslant \alpha - d\alpha, \quad (5.4)$$

where

$$\alpha = \sin^{-1}\left(\frac{v}{V_0}\right) \quad (5.5)$$

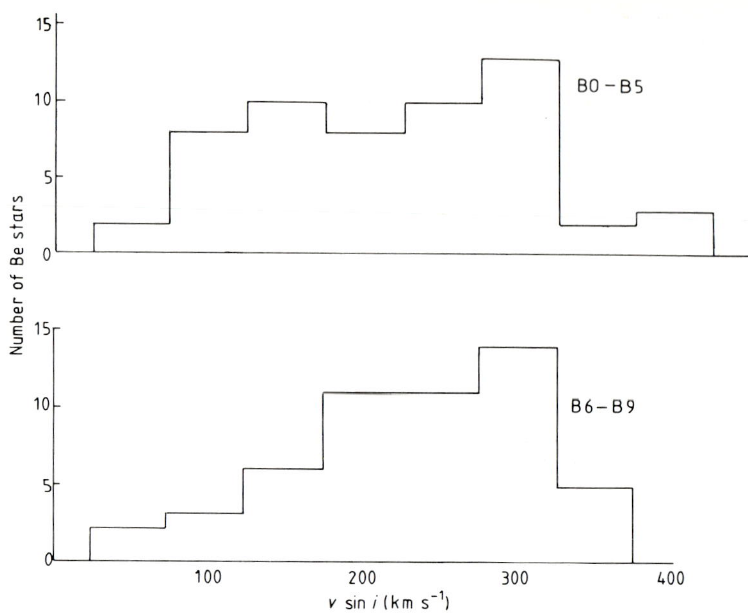

Figure 5.7. Distribution of $v \sin i$ in Be stars. (Reproduced from Slettebak 1976 by permission.)

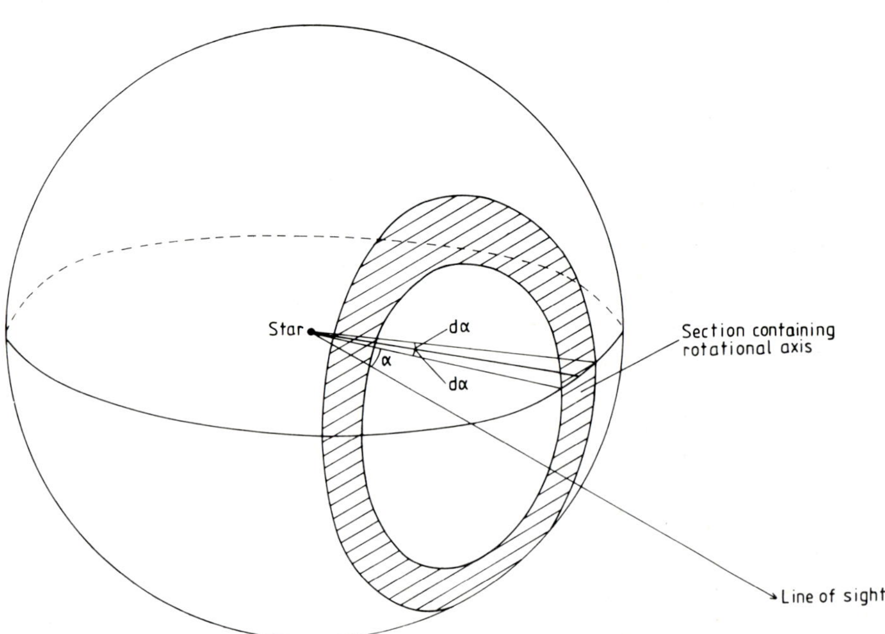

Figure 5.8. Distribution of stellar rotational axes.

and

$$\mathrm{d}\alpha = \sin^{-1}\left(\frac{\mathrm{d}v}{(V_0^2 - v^2)^{1/2}}\right). \tag{5.6}$$

The proportion of the sphere occupied by this annulus is

$$\frac{(v/V_0)\,\mathrm{d}(v/V_0)}{2\,[1-(v/V_0)^2]^{1/2}}. \tag{5.7}$$

Hence for a large number of stars whose axes are randomly distributed, the number whose observed rotational velocity will lie in the range $v \pm \mathrm{d}v$, N_v, will be

$$N_v \propto \frac{(v/V_0)\,\mathrm{d}(v/V_0)}{[1-(v/V_0)^2]^{1/2}}. \tag{5.8}$$

This is a very sharply peaked function, and its shape is plotted in figure 5.9. Since the observed distributions are likely to contain a range of actual equatorial rotational velocities, such a sharp peak and cut-off are not to be expected. None the less the distribution of the later spectral types (figure 5.7) does resemble the theoretical curve, and while the curve for the earlier spectral types is not so clear cut, there is a tendency to peak at the higher velocities. Thus it is at least consistent with the observations that the Be stars are all rotating rapidly, with equatorial rotational velocities near 400 to 450 km s^{-1}.

A velocity as high as that suggested above is close to the break-up velocity of the stars (figure 5.10). But it is no longer clear that they are rotating at the

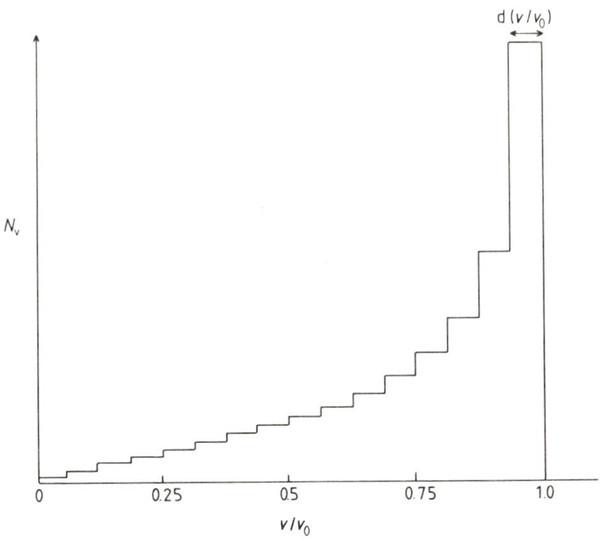

Figure 5.9. Distribution of the observed rotational velocities of a randomly ordered uniformly rotating set of stars.

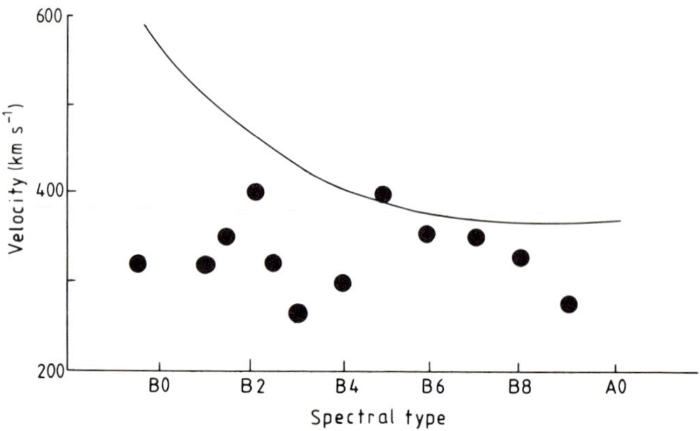

Figure 5.10. Plot of the maximum observed values of $v \sin i$ (dots), and the break-up velocity of main sequence stellar models (full curve). (Reproduced from Slettebak 1976 by permission.)

break-up velocity, even though this hypothesis was generally accepted until fairly recently. Be stars occupy a region just above the main sequence, which might suggest that they are evolving towards the giant region, and so would have a lower critical velocity. However, Roxburgh and others (Collins and Sonneborn 1977, Roxburgh and Strittmatter 1965, Warren 1976) have shown that rapid rotation moves the main sequence by about one magnitude, giving a similar effect to evolution. Thus rotation, while very rapid, is probably lower than the critical velocity. The surrounding gaseous envelope which produces the emission lines therefore probably requires additional processes such as radiation pressure, surface activity, binary interactions etc for its formation. Such processes are probably enabled to expel material from the star by the reduced effective gravity in the equatorial regions arising from the rotation.

5.4. Emission Lines

The commonest, and usually the most intense, emission line occurs at Hα. In some Be stars this may be the only emission line. Normally, however, emission lines occur throughout the spectrum. In the infrared, the Paschen series, O I 844.6 and 777.2, and Ca II 849.8 are frequently in emission (figure 5.11). In the optical region, the Balmer series to about H 10, He I 501.5, 587.5, 667.8, Mg II 448.1, and numerous Fe II lines are found, while in the ultraviolet the C IV lines at 155 nm often exhibit emission, sometimes as part of a P Cygni line profile. The profiles of other lines are of types '*d*', '*e*' and '*f*' of figure 5.1 for many of the Balmer lines. The Paschen lines are usually straight emission lines (double or single) without underlying absorption. The He I, Fe II and Mg II lines etc have similar profiles to the Paschen lines. In the ultraviolet the lines usually have P Cygni profiles.

The emission lines vary in strength and shape, as has been mentioned earlier. Sometimes they disappear altogether. At other times they appear in a previously

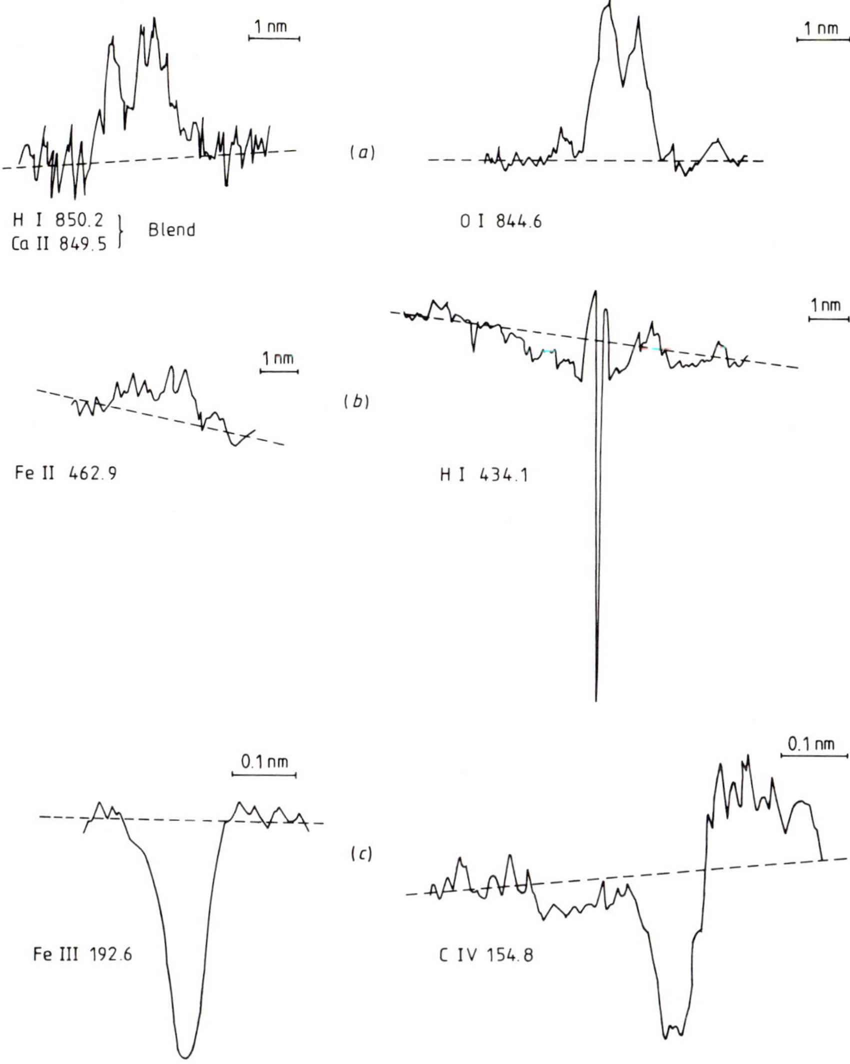

Figure 5.11. (*a*) Infrared, (*b*) visual and (*c*) ultraviolet line profiles in ϕ Per.

normal star. The Be phenomenon therefore appears, in at least some stars, to be a phase which varies markedly in intensity. Many early stars, which at present are thought to be normal, may therefore merely be in the quiescent phase of their cycle.

Rapid variations in the emission lines have been noted by several observers (Bahng 1976, Doazan 1976, Hutchings 1969, Swings 1973) (figure 5.12) but it is not clear how much of the variation may be attributable to noise in the detecting system. Doazan (1976) finds no correlation between the Hβ and the Hγ variations, while Lacy (1977) found no significant variations at all. It is better for the moment to treat observations of rapid emission line variations with reserve.

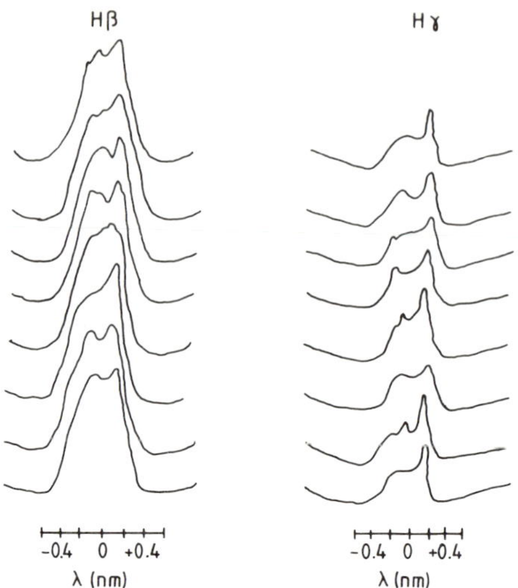

Figure 5.12. Profiles of Hβ and Hγ in HD 5394. (Exposures of 4 to 11 minutes at approximately 12 minute intervals.) (Reproduced from Doazan 1976 by permission.)

It is universally agreed that the emission lines are produced in an extensive rarefied gaseous envelope which surrounds the star and which is probably equatorially concentrated. (The envelope models are discussed in more detail later.) The emission lines generally arise through dilute radiation in the extended envelope allowing the Rosseland cycle to operate (chapter 1). At least one line however, the O I 844.6 line, probably arises through selective fluorescence with the O I 102.577 absorbing Lyβ at 102.572 nm. A plot of the equivalent widths of Hα and O I 844.6 (figure 5.13) shows a linear relationship, since both depend on Lyβ for their initial excitation. One consequence is that O I 130.2 should also be in emission whenever O I 844.6 is in emission (figure 5.14). However, in ϕ Per this does not happen (figure 5.15) (Kitchin 1982), only a much narrower interstellar absorption

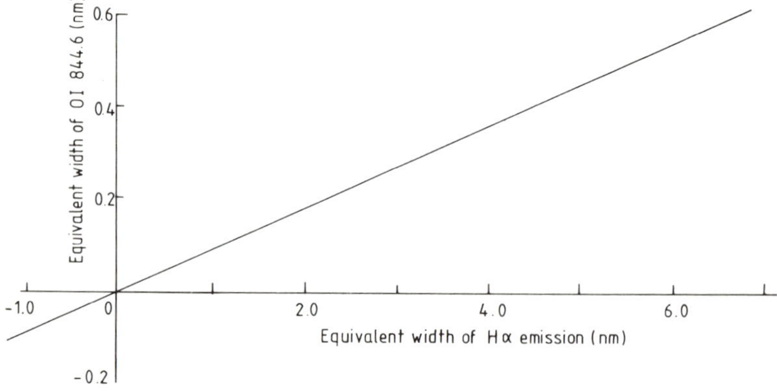

Figure 5.13. Variation of the O I 844.6 emission intensity with that of Hα.

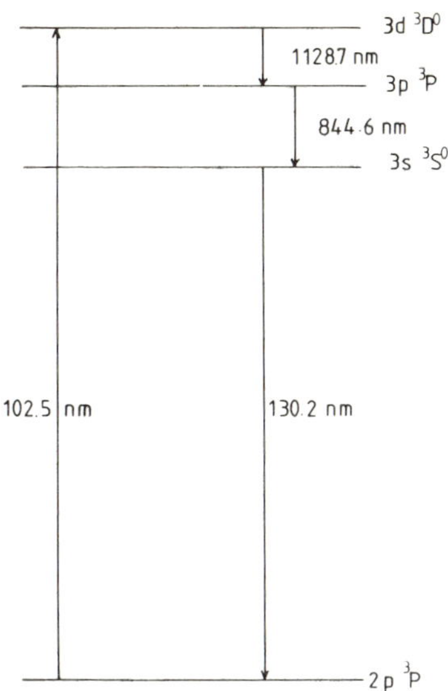

Figure 5.14. Partial Grotrian diagram of O I.

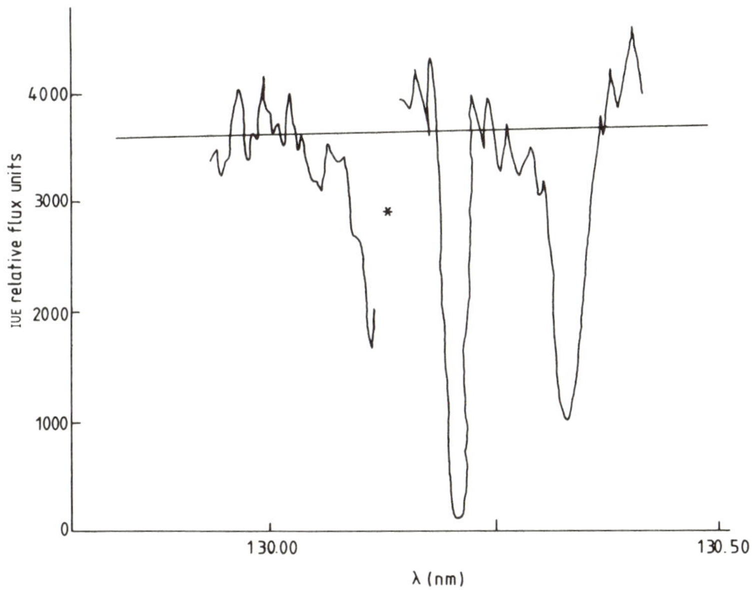

Figure 5.15. The O I 130.2 line in ϕ Per. (The asterisk marks a gap in the tracing due to a camera réseau mark.) (Reproduced from Kitchin 1982 by permission.)

component being visible, and so the selective fluorescence process remains to be completely confirmed in this case.

The separations of the components of double Balmer emission lines varies with line number (figure 5.16). This indicates that the early Balmer lines are optically thick, and so are formed in low-velocity regions near the outside of the envelope.

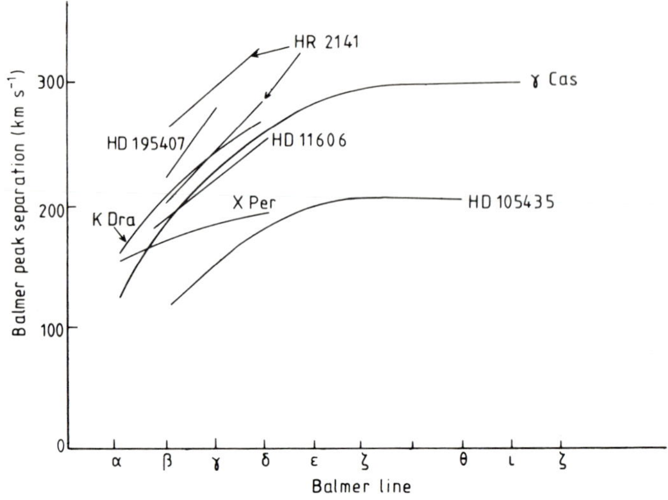

Figure 5.16. Balmer emission line peak separation. (Reproduced from Hutchings 1976b by permission.)

The lines become optically thin around H6 to H8 and the separation then reflects the photospheric rotation (the actual value of $v \sin i$ will depend on the model of the system).

5.5. Absorption Lines

The broad absorption lines are usually attributed to photospheric absorption in the star. They show no significant differences (apart from their breadth) from the absorption lines of a normal B type star.

The narrow absorption lines show many of the behavioural characteristics of the emission lines, and often their changes correlate with the E/C variations of the emission lines. Their widths are much narrower than those of the emission lines (typically the widths are 100 km s^{-1} for the narrow absorption lines, 200 km s^{-1} for the emission lines and 400 km s^{-1} for the broad absorption lines). In the visible and infrared they are usually present as absorption cores to the emission or broad absorption lines. The majority of the lines in the ultraviolet are simple narrow absorption lines (Kitchin 1982). These lines are likely to be formed far out from the star, perhaps 20 to 50 stellar radii out into the envelope (Kitchin 1973b).

5.6. Photometry

The absolute visual magnitude of the Be stars ranges from about -3^{m} at B2e to about 0^{m} at B9e. These figures place the Be stars above the main sequence by about one magnitude (figure 5.17). This position off the main sequence has been attributed to the evolution of the stars away from it, with an age of perhaps 3×10^{7} years (Mendoza 1958). An evolutionary position at the core contraction stage has

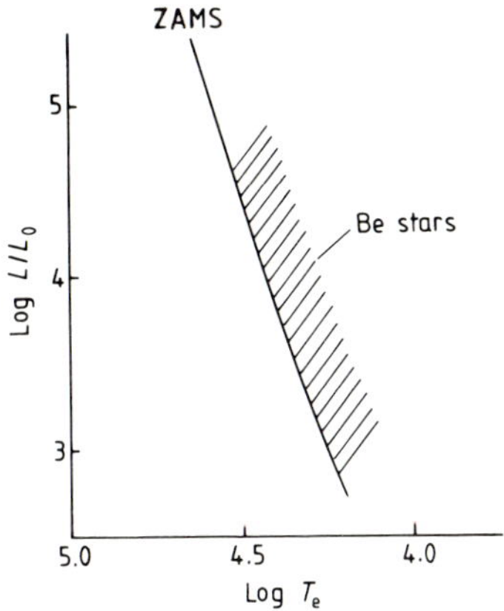

Figure 5.17. Position of the Be stars on the H–R diagram.

been suggested for the Be stars in clusters to explain their brighter than average magnitudes (Lloyd-Evans 1980). However, as mentioned earlier, rapid rotation probably affects the appearance of the star in a similar way to evolution (figure 5.18), and so provides a more probable explanation for their absolute magnitudes.

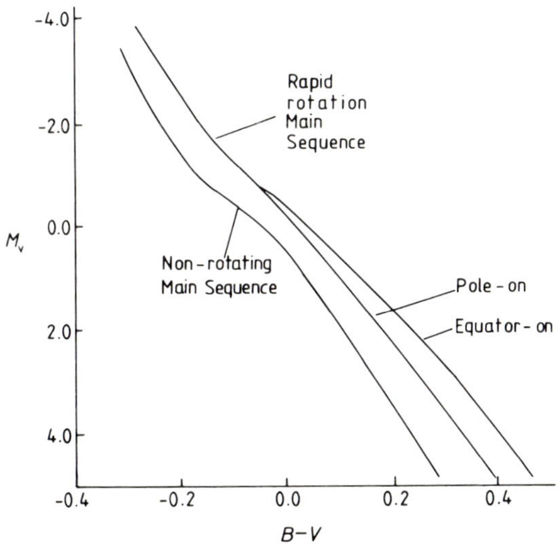

Figure 5.18. Effect of rapid rotation on the main sequence. (Reproduced from Roxburgh and Strittmatter 1965 by permission.)

Variations in magnitude of a few tenths of a magnitude, and with timescales ranging from minutes to years, occur in a majority of Be stars. Periodic variations in HD 217050 by 0.2^m in 0.8 days may be attributable to rotation (Walker 1953). Larger periodic variations are found in other stars which are probably due to their binary nature. For example the light curve of HD 187399 is shown in figure 5.19, where the variations may be due to obscuration by matter being exchanged

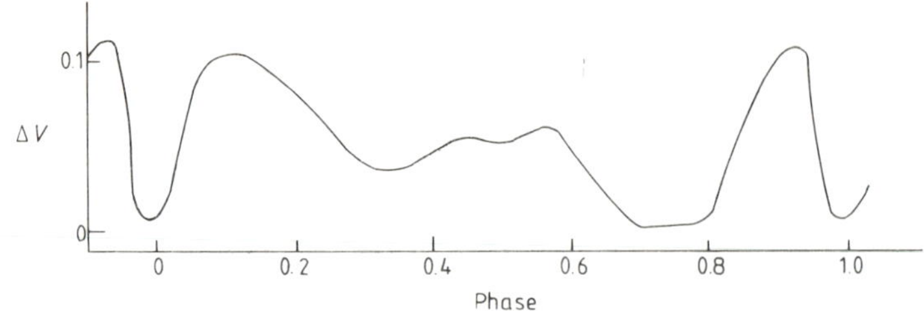

Figure 5.19. Light curve of HD 18733. (Reproduced from Harmanec and Kriz 1976 by permission.)

between the two stars (Harmanec and Kriz 1976), or to the non-sphericity of the stars (Hutchings 1976b). Larger amplitude variations occur over longer periods. HD 23862 (Pleione) has varied by almost a magnitude (figure 5.20), and its decreases in brightness are associated with increases in its extended envelope phenomena.

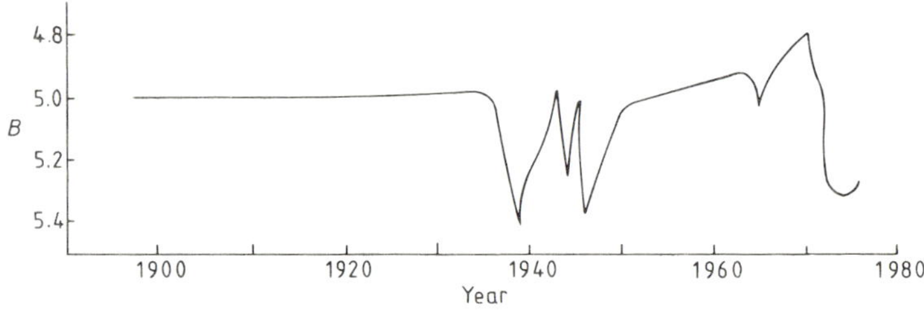

Figure 5.20. Light curve of HD 23862. (Reproduced from Sharov and Lyuty 1976 by permission.)

Strong infrared excesses occur in some extreme Be stars, which may extend well into the visible spectrum (figure 5.21), and which may be interpreted as H^- free-bound emission in the envelopes (Schild *et al* 1974). The fit to the theoretical curve, however, is not very good, and so other processes may need to be invoked. HD 45677 emits some 80% of its energy in the infrared, for example, probably from re-emission by dust particles (Apruzese 1974, Sitko and Savage 1980, Swings and Allen 1971). Most more normal Be stars also have an infrared excess, although not as pronounced as that of the Bex stars (figure 5.22). This is attributed to free-free and free-bound emission (Gerhz *et al* 1974, Savage *et al* 1978, Whittet and Breda 1980). In γ Cas the infrared excess leads to an estimate of the mass loss rate of a few times $10^{-7} M_\odot \, a^{-1}$ (Ferrari-Toniolo *et al* 1978), and a shell temperature of 18 000 K (Savage *et al* 1978). The variable Be star X Per only exhibits an infrared excess when most luminous, and this behaviour may perhaps be linked with the extension of the circumstellar envelope proposed to explain the optical variability (Persi *et al* 1977). In this case a mass loss rate of $10^{-9} M_\odot \, a^{-1}$ is indicated. The infrared excess declines steeply at long wavelengths, and there is a correlation

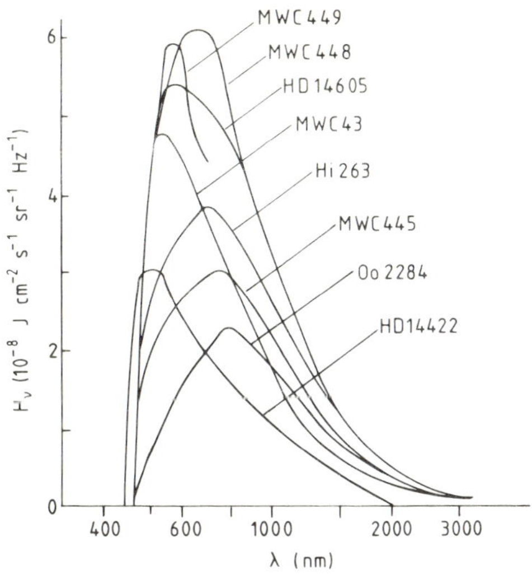

Figure 5.21. Infrared excesses of Be stars in h and χ Persei. (Reproduced from Schild 1976 by permission.)

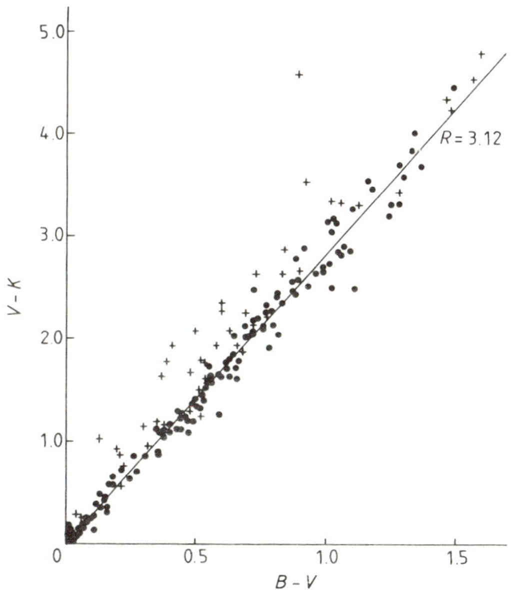

Figure 5.22. *V-K/B-V* diagram for early type stars. (Emission line stars are indicated by crosses.) (Reproduced from Whittel and Breda 1980 by permission.)

between this 'turn-over' and the presence of large optical polarisation. Both of these effects are attributable to a disc-shaped envelope with the density proportional to r^{-2} to r^{-3}, an electron scattering optical depth of 0.5 to 1.3, and a continuum optical depth of 0.1 (Hartmann 1978b).

The earliest Be stars have a negative Balmer discontinuity (figure 5.23), implying that scattering is not the dominant source of opacity in the envelope.

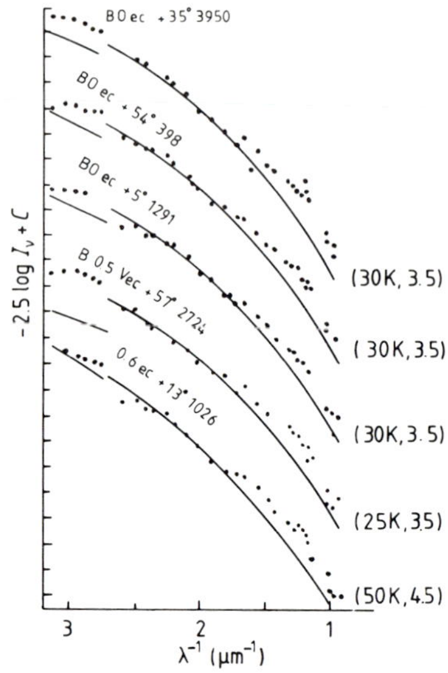

Figure 5.23. Balmer discontinuities in early Be stars fitted to appropriate models (temperatures and $\log g$ are listed at the side of each model). (Reproduced from Schild 1976 by permission.)

5.7. Ultraviolet Spectrum

The ultraviolet spectrum contains the resonance and other ground state transitions of most of the commoner elements and so, potentially, contains a wealth of information compared with the relatively line-free optical spectra of early stars. In Be stars the work to date has concentrated on stellar winds and mass loss, although some work on abundances, surface gravity and effective temperature has also been undertaken.

Mass loss rates are generally around $10^{-9} M_\odot\ a^{-1}$ (Hammerschlag-Hensberg *et al* 1980, Kitchin 1982, Lamers and Rogerson 1978, Marlborough and Cowley 1974, Marlborough and Snow 1976) at least in the earlier Be stars although values as high as $10^{-7} M_\odot\ a^{-1}$ have been found (Hutchings 1970a), while the metastable Fe III lines at 206.1 to 207.9 nm in ϕ Per suggest only $5 \times 10^{-11} M_\odot\ a^{-1}$ (Bruhweiler *et al* 1978) (the resonance lines of this star, however, sugggest a more normal rate of $10^{-9} M_\odot\ a^{-1}$ (Hammerschlag-Hensberg *et al* 1980)). There is a correlation between $v \sin i$ and the presence of signs of mass loss in the spectrum of a Be star (Marlborough 1977, Marlborough and Snow 1976) (figure 5.24), and mass loss occurs in later spectral types than for normal B type stars. Both these effects arise from the reduced effective gravity in the equatorial region of a Be star, due to their rapid rotation. If the mass flow is preferentially equatorial, then the correlation with $v \sin i$ indicates that low $v \sin i$ stars are indeed seen pole-on as has long been

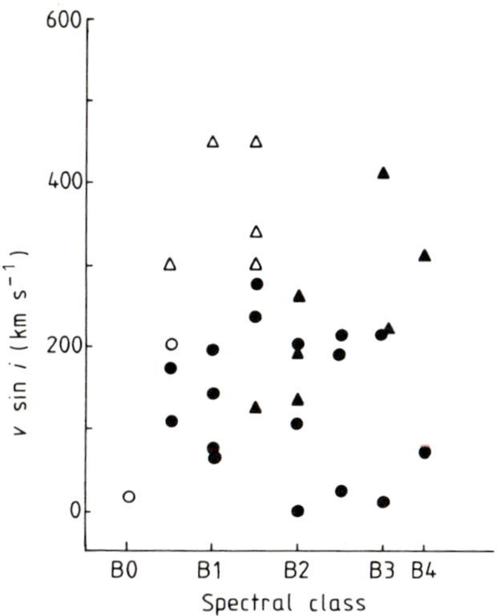

Figure 5.24. Correlation of high velocity mass loss with $v \sin i$. △ Be star showing mass loss, ▲ Be star not showing mass loss, ○ B star showing mass loss, ● B star not showing mass loss. (Reproduced from Marlborough and Snow 1976 by permission.)

suspected. There is also a correlation of expansion velocity and degree of ionisation, suggesting a stellar wind in which both velocity and degree of ionisation increase outwards (Snow *et al* 1979) for some Be stars. The wind is very variable and results in major changes in the spectrum. In particular the narrow absorption components of C IV, N V and Si IV in γ Cas, which have velocities of about -1500 km s^{-1} appear and disappear with a timescale of a few days, and the ultraviolet lines in 59 Cyg lead to highly variable mass loss rates (Doazan *et al* 1980).

There is a strong correlation between the Mg II lines at 280 nm and the Fe II lines near 254 nm due to ultraviolet multiplet number 158 (Ringuelet 1980) (figure 5.25). These Fe II lines seem likely to arise by collisional excitation and to be formed over the same region as the Mg II lines. They may well, therefore, be useful in studying the outer layers of the star's atmospheres. Features at 172 and 192 nm due to Al II and Fe III are also found in the spectra of supergiants and are general indicators of extended atmospheres (Heap 1977b). In at least two stars, HD 45677 (B2 IVe) and HD 50138 (B8e), there is a comparative lack of radiation in their ultraviolet spectra which is approximately balanced in energy terms by their infrared excesses (Savage *et al* 1978), which is further evidence of the importance of thermal re-emission by dust in some Be stars. Many of the early Be stars have a deficiency in their energy near 210 nm, which is unidentified but whose size correlates with other indicators of the presence and extent of the envelope (Beeckmans and Hubert-Delplace 1980).

5.8. X-ray Observations

γ Cas and X Per have been identified with x-ray sources (Bradt *et al* 1977, Mason *et al* 1976). X Per is variable and its associated x-ray source, 3U 0352 + 30, has

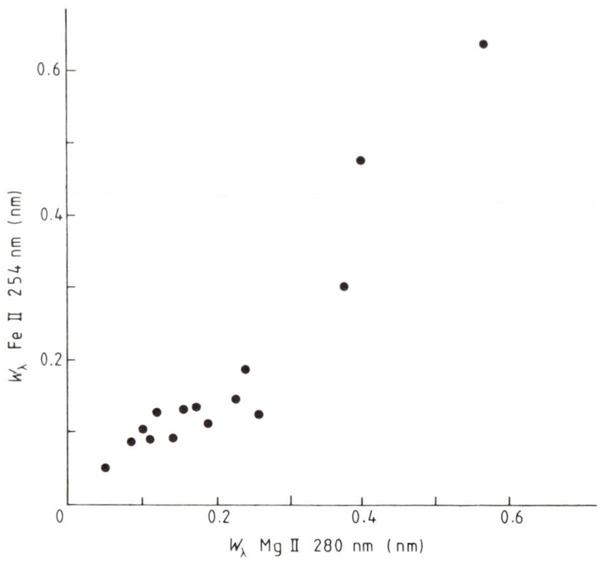

Figure 5.25. Correlation between the Fe II lines near 254 nm and the Mg II lines near 280 nm in B and Be stars. (Reproduced from Ringuelet 1980 by permission.)

periods of 13.9 minutes and 22 hours. The x-ray variability may correlate with the infrared variations but there is no correlation with the optical changes (Frontera *et al* 1979, Hutchings 1977, Margon *et al* 1976). γ Cas may also have variable x-ray emission.

The weak x-ray pulsar GX304-1 (also known as 4U1258−61) may be identifiable with a fifteenth magnitude Be star (Mason *et al* 1978b, Parkes *et al* 1980). Mass loss from the Be star may be the mechanism which results in accretion on to the x-ray companion, and produces the x-rays. Another possible identification is the O9.7 IIe star HDE 245770 with the transient x-ray source A 0535 + 26 (Liller 1975, Murdin 1975). At its peak this transient source was twice as bright as the Crab Nebula in the 1 to 6 keV region and varied with a period of 104 seconds (Rappaport *et al* 1976, 1979). SMC X2 may perhaps be identified with a faint Be star (Murdin *et al* 1979), and HD 28497 (B1.5 Ve) may have been detected in the soft x-ray region (Long and White 1980).

5.9. Radio Observations

Normal Be stars have not been detected in the radio region, and upper limits of 20 mJy have been set for their radio emission (Purton 1976, Schwartz and Spencer 1977). (χ Dra may just have been detected at 5 GHz with an emission of 20 ± 15 mJy (Tarasco *et al* 1970).) A few peculiar objects which may be related to Be stars have been detected. For example MWC 349 whose emission is shown in figure 5.26. The spectral index, α (defined by $S \propto \nu^{\alpha}$) is 0.7, which leads to an

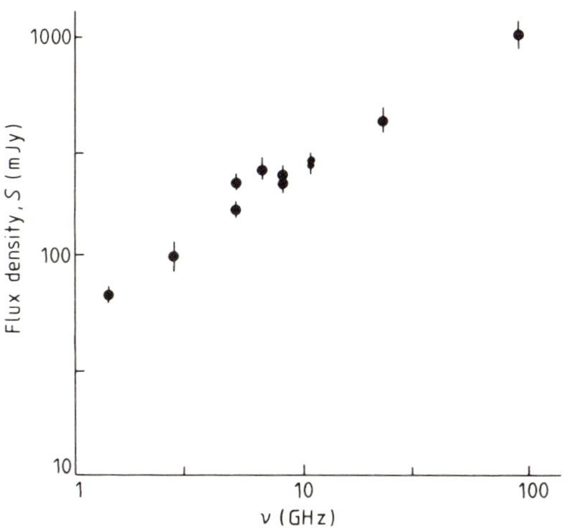

Figure 5.26. Radio emission of MWC 349. (Reproduced from Purton 1976 by permission.)

electron density distribution of (Purton 1976)

$$N_e = 3.8 \times 10^{40} r^{-2.1} \quad \text{m}^{-3}, \tag{5.9}$$

where r is the radius in metres, and to inner and outer radii of the envelope of 10^{13} and 10^{15} m, respectively. This value for the spectral index almost matches the theoretical one of 0.6 which is to be expected for an envelope formed by mass loss whose far infrared and radio emission arises from free–free radiation (Wright and Barlow 1975). The size of the envelope suggests that this object may be better classed as a small planetary nebula than as a star with an extended envelope.

5.10. Polarimetry

Be stars are strongly concentrated towards the Galactic plane (Wackerling 1970), and so the initial problem in any study of their polarisation is to separate out the interstellar component. This may be accomplished in several ways (Coyne 1976):

(*a*) from the variability of the polarisation with time (the interstellar component being assumed constant),

(*b*) from a peculiar variation (i.e. radically different from the normal interstellar variation) of the degree of polarisation with wavelength,

(*c*) from the variation of polarisation across emission lines.

Most of these methods are used in practice, and the resulting polarisation curves are shown in figure 5.27 for the continuum, and in figure 5.28 for an individual line. There is a more rapid decrease of polarisation towards the ultraviolet than for interstellar polarisation, a minimum on the short-wavelength side of the Paschen limit, and a further decrease (not shown) at wavelengths longer than 1 μm.

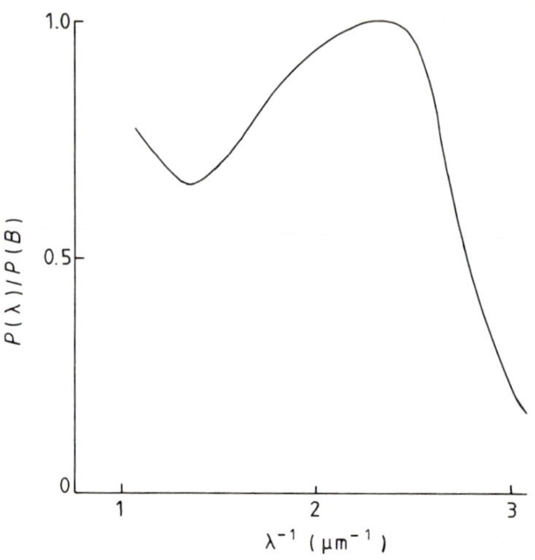

Figure 5.27. Normalised polarisation curve for P Car and α Ara. (Reproduced from Coyne 1976 by permission.)

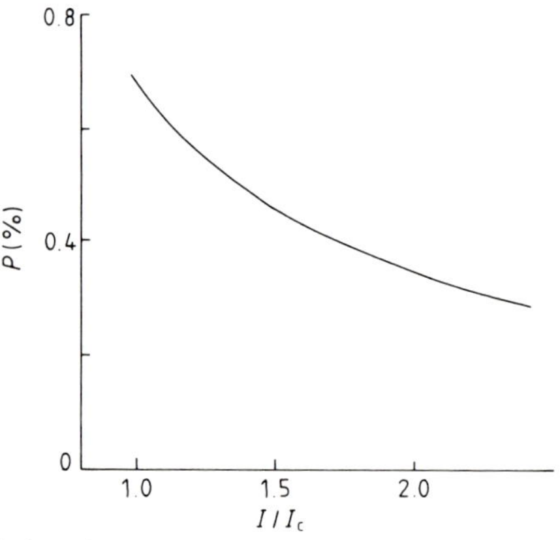

Figure 5.28. Variation in the polarisation across Hα in 48 Lib. (Reproduced from Coyne 1976 by permission.)

However, considerable variations from this pattern occur for individual stars. Furthermore the total polarisation depends on $v \sin i$; there is little or no polarisation found in those stars which have low values of $v \sin i$, and the largest polarisation is found in stars with the largest values of $v \sin i$ (Poeckert and Marlborough 1976). The polarisation curves can be produced by electron scattering modified by hydrogen absorption in an equatorially flattened envelope. The path-length in the envelope is from three to ten stellar radii, the electron temperature about 10 000 K and the electron density about $10^{18}\,\mathrm{m}^{-3}$ (Coyne 1976, Haisch and Cassinelli 1976, Jones 1979).

The polarisations of most Be stars change with timescales from a few hours to a few days (Piirola 1979), but so far no correlation has been found with any of the other variable features.

The variation of the polarisation across the emission lines (figure 5.28) shows a decrease from the wings to the centre of the line. In a few cases there is an increase in polarisation by a very small amount right at the centre of the line (McLean *et al* 1979a). The emission line flux is thus less polarised than the continuum flux. The emission (especially of Hα) must therefore arise from a region where the electron scattering optical depth is less than for the regions producing the continuum, i.e. the outermost parts of the envelope (George and Coyne 1974, McLean 1979, McLean and Clarke 1979, Poeckert 1975).

5.11. Models

There is general agreement that a Be star is a B type star surrounded by an extensive rarefied gaseous envelope. Where disagreement occurs it is over the origin and supporting mechanism(s) of the envelope, and to some extent over its size and shape.

The shape in most models is equatorially concentrated, with an outer radius of 2 to 20 stellar radii (10^{10} to 10^{11} m) (Kitchin 1970b, 1973a, b), and with a thickness of at least one stellar diameter. The material may be in the form of a complete disc, or there may be an absence of material in the centre producing a ring-shaped envelope. In such a case, the ring may be elliptical rather than circular in shape. In the binary models, the material producing the emission is streaming between the stars, while in the Herbig Ae and Be stars (see the final section of this chapter) it is the remnant of the contracting interstellar gas cloud, and so may be closer to a sphere in shape than to a disc. The observational constraints on the models are ambiguous, and do not at present lead to clear and definitive tests to distinguish between the various hypotheses (Poeckert and Marlborough 1978). Each of the competing theories is therefore discussed in detail below.

5.11.1. Elliptical Ring Model

This was first suggested by Struve (1931) and developed more quantitatively by Huang (1973, 1976). The ring particles orbit the star in highly elliptical Keplerian orbits concentrated towards the equatorial plane. The particle density is greatest at apoastron and least at periastron due to the varying orbital velocity. The ring as a whole will rotate due to tidal action of the stellar equatorial bulge, and due to interaction with continually escaping material (Huang 1977, 1978) (figure 5.29). Such a model fits the observations of some stars (e.g. 105 Tau, HD 20336, 25 Ori, β' Mon) quite well (figure 5.30) for rings with radii of three to four stellar radii, and eccentricities of 0.2 to 0.3.

The model describes the V/R variations adequately, however, it suffers from a number of drawbacks. The ring has to be narrow, and it is therefore difficult to produce the central absorption lines unless the inclination of the star is near 90°. For stars in which the inclination is significantly different from 90°, an extensive

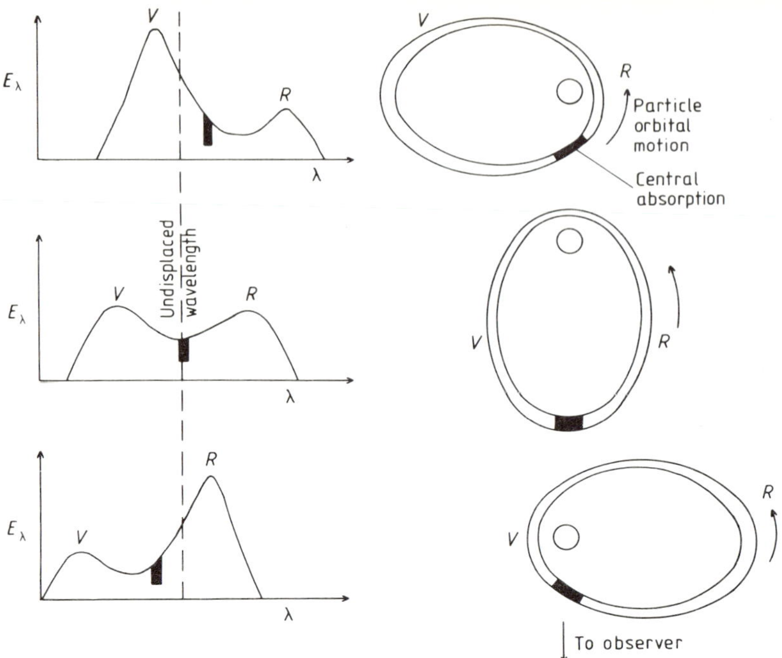

Figure 5.29. Elliptical ring model for Be stars showing the V/R ratios for differing orientations of the ring. (Reproduced from Huang 1975 courtesy of *Sky and Telescope*.)

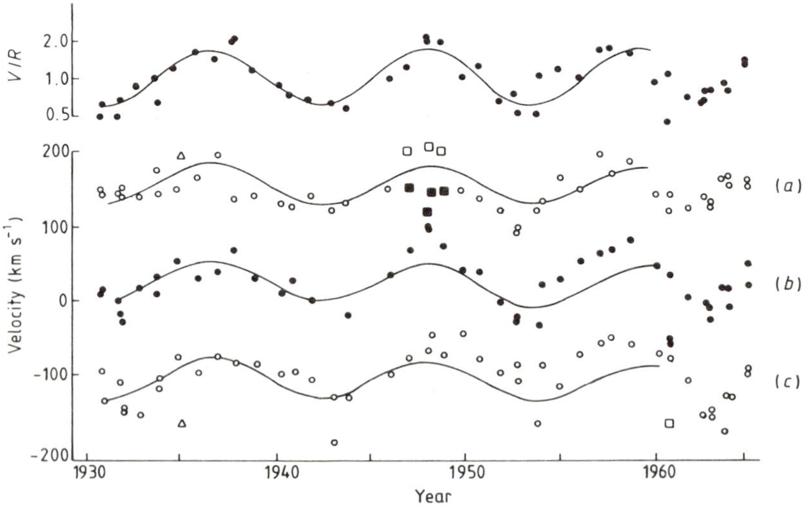

Figure 5.30. Comparison of observed data for 105 Tau and the predictions of an elliptical ring model. (*a*) Velocity of the red emission edge, (*b*) velocity of the central absorption, (*c*) velocity of the violet emission edge. (Reproduced from Huang 1973 by permission.)

absorbing region outside the star and ring has to be postulated, but this will then produce a central absorption at a constant wavelength. It is also not clear how the ring of material would form. While radiation pressure may be able to eject material from the regions with rotationally lowered effective gravities, such material would not then have sufficient angular momentum to orbit the star. If magnetic or turbulent viscosity were sufficient to transfer angular momentum from the star to the ejected material, then such forces would also be strong enough to modify the particle's Keplerian orbits. Huang suggests that the ring forms from only that small fraction of the ejecta which does have a high enough angular momentum, while the remainder falls back on to the star (or escapes completely).

5.11.2. *Disc Models*

These models bear some resemblance to the elliptical ring, but they are assumed to be circular and wide. They may even extend down to the stellar surface (contact discs). Velocity functions ranging from Keplerian orbits to radial expansion with a variety of acceleration functions have been postulated by various authors. The shapes may be flat discs, lenticular, or increase in thickness away from the star. The size may be from one to fifty stellar radii for the outer edge, and from one to five stellar radii for the inner edge. The thickness may range from a fraction to several stellar radii. This range of unfixed parameters ensures that it is possible to get a good fit of theoretical line profiles to most of the observed line profiles. For example the profile of Hγ in γ Cas for the schematic model shown in figure 5.31 is compared with the observed profile in figure 5.32 (Hutchings 1970a). The disc model, however, cannot explain a value of $V/R > 1.0$, except by material falling on to the star. There is no evidence for infall of material in Be stars, and even if it were to occur, it is difficult to see how its velocity could exceed the local speed of

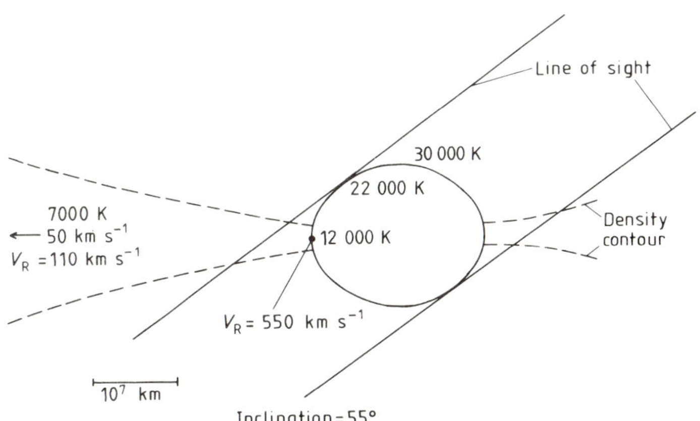

Figure 5.31. Schematic model of γ Cas. (Reproduced from Hutchings 1970a by permission.)

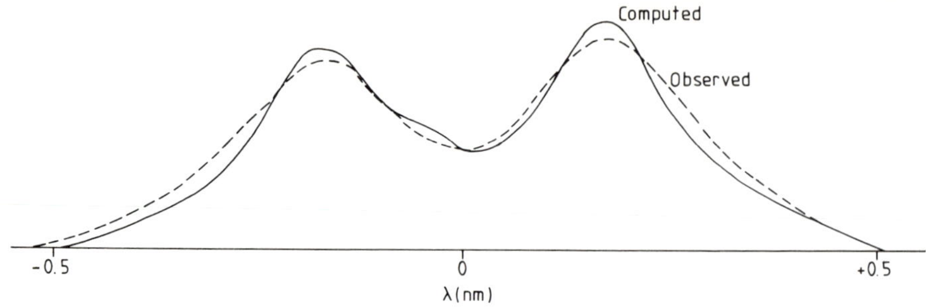

Figure 5.32. Observed line profile of Hγ in γ Cas compared with the theoretical profile for the model shown in figure 5.31. (Reproduced from Hutchings 1970a by permission.)

sound. Since the speed of sound is given by

$$v_s = \left(\frac{\gamma k T}{\mu m_H}\right)^{1/2}, \qquad (5.10)$$

the expected velocities should be less than 10 to 30 km s^{-1}. Values of V/R significantly greater than one would require velocities one or two orders of magnitude greater than these for their production. Values of V/R less than one require expansion in the envelope. Evidence for mass loss and stellar winds has been reviewed earlier, and the observed flows seem adequate to explain the observation of $V/R < 1.0$. There is no obvious reason on these models why V/R should vary cyclically. It is also difficult to find a supportive mechanism (except for those models which assume Keplerian orbits, and these encounter the same problems as the elliptical ring model). Magnetic fields of several tens of mT would be suitable, but even with the width of Be star lines such fields should be detectable. Furthermore there are only limited possibilities for steady-state solutions of the coupling of a magnetic field to the envelope (Limber 1976). (The instability of the situation could, however, provide a mechanism for the short-term variations found in Be stars.) In the stellar wind models, radiation pressure is the supportive mechanism of the envelope. However, the effective gravity, and hence also the temperature by Von Zeipel's theorem (Eddington 1959), is least in the equatorial regions, so that the radiation pressure will be lowest near the equator. A Be star of type B1e will range from about B3 or B4 at the equator to about O9 at the poles. In slowly rotating stars significant mass loss only occurs for stars earlier than spectral type B1 (Snow and Morton 1976). Rotation does not lower the effective gravity by large fractions until the star is rotating near its break-up velocity; the effective gravity is only halved when the star is rotating at 75% of its break-up velocity for example. Thus it is not clear that radiation pressure alone will be sufficient to initiate equatorial mass loss, and hence formation of a disc. Surface activity (prominences, flares etc) may be required in addition in these models. A steady-state mass loss model incorporating gravitation, pressure and magnetic fields, has been investigated by Limber (1974), but this has no solutions for most values of its input parameters. Other stellar wind models have been suggested by Marlborough and others (Marlborough 1969, 1970, Marlborough and Cowley 1974, Marlborough and Roy 1971), and by Doazan (1965), but these require further *ad hoc* assumptions to explain the *V/R* and other variations. The suggestion has been made that in Bex stars the mass loss may be initiated or enhanced by core concentration occurring at the end of the core hydrogen burning phase (Hutchings 1976b, Schild *et al* 1976), but as discussed earlier the evolutionary position of Be stars is complicated by their rotation, and this suggestion is not universally accepted.

5.11.3. Binary Models

A number of Be stars are known to be members of binary systems. Most of the Be phenomena can qualitatively be reproduced by mass streaming in a binary (figure 5.33). The model postulates an essentially unevolved A or B star with a small radius, and a low-mass secondary which fills its Roche lobe. Mass exchange occurs to the primary through the inner Lagrangian point, and may also occur at L_2 and L_3. The emission line producing envelope forms inside the Roche lobe of the primary. Periodic variations are easy to understand on this model, and irregular variations on all timescales are likely to be common in the gas streams between the stars. A major objection, at least to the application of this model to a majority of Be stars, is the lack of eclipsing binary systems containing Be stars. With mass ratios between three and ten, then between 20% and 30% of all Be stars should show detectable eclipses (Plavec 1976). It therefore seems probable that this model can only explain a small fraction of Be stars.

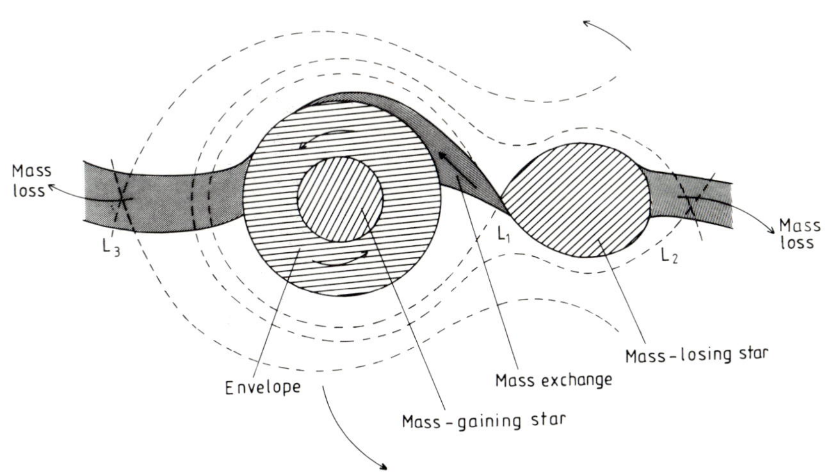

Figure 5.33. Schematic binary model of a Be star. (Reproduced from Kriz and Harmanec 1975 by permission.)

5.12. Herbig Ae and Be Stars

Herbig (1960) suggested in 1960 that there might be high-mass analogues of the T Tauri stars. These would be pre-main sequence objects, and would be recognisable by:

(*a*) emission lines in their spectra,
(*b*) association with highly obscured regions,
(*c*) illumination of bright nearby nebulosity.

Herbig identified 26 such stars, and subsequent study of them has largely confirmed his original hypothesis. The emission component of Hβ in HD 97048 for example can clearly be distinguished in figure 5.34. The stars are found to be strong infrared sources. The infrared spectrum of HD 97048 is shown in figure 5.35, and contains unidentified features at 3.28, 3.41 and 3.50 μm. (The first two of these features have also been found in Galactic nebulae (Blades and Whittet 1980).) At longer wavelengths the spectra are largely featureless, except for the possible existence of very weak absorption near 10 μm which may be due to silicates (Cohen 1980). HD 97048 is associated with the chamaeleon dark cloud, and has an infrared signature characteristic of a dust shell at about 800 K (Glass 1979). In general the infrared data require circumstellar dust shells (perhaps, but not necessarily, disc-shaped) up to 10^{12} m in radius and with temperatures between 500 and 1000 K (Gillett and Stein 1971). The lifetime of a dust particle near a hot star is only 10^4 to 10^5 years (Mathews 1969) and so these dust shells must either be replenished from outside, or be in the process of dispersal following the recent formation of the star. The latter seems the more probable, and a mass loss rate of $2.3 \times 10^{-9} M_{\odot}\, a^{-1}$ is estimated for HD 200775, the Herbig Be star associated with NGC 7023 (Altamore *et al* 1980).

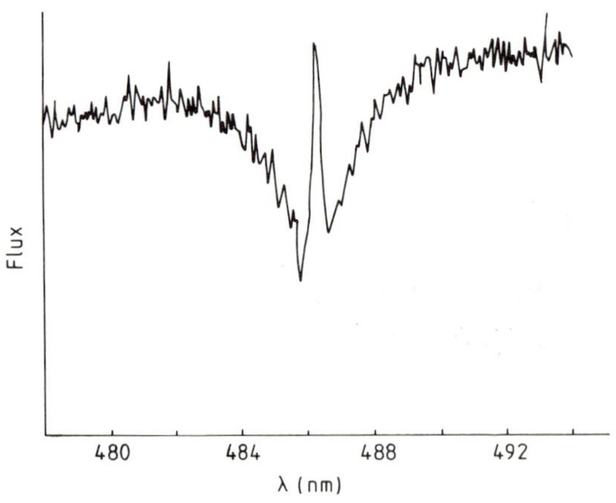

Figure 5.34. Hβ profile in HD 97048. (Reproduced from Blades and Whittet 1980 by permission.)

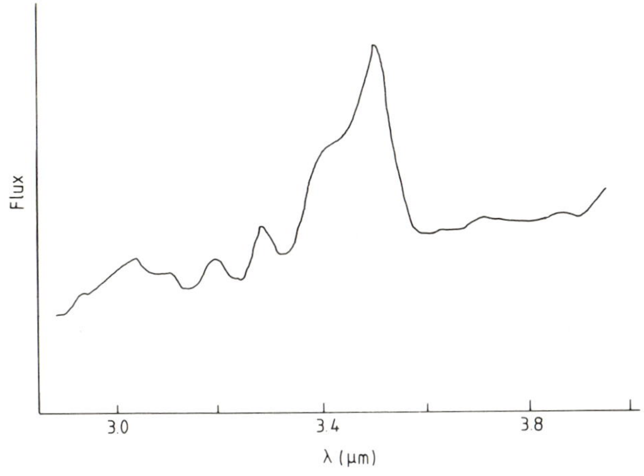

Figure 5.35. Near infrared spectra of HD 97048. (Reproduced from Blades and Whittett 1980 by permission.)

The polarisation curves of Herbig Be stars differ from those of normal Be stars (figure 5.36) with the maximum occurring in the red instead of in the blue part of the spectrum (Vrba *et al* 1979). In some cases there is also an unexplained peak at

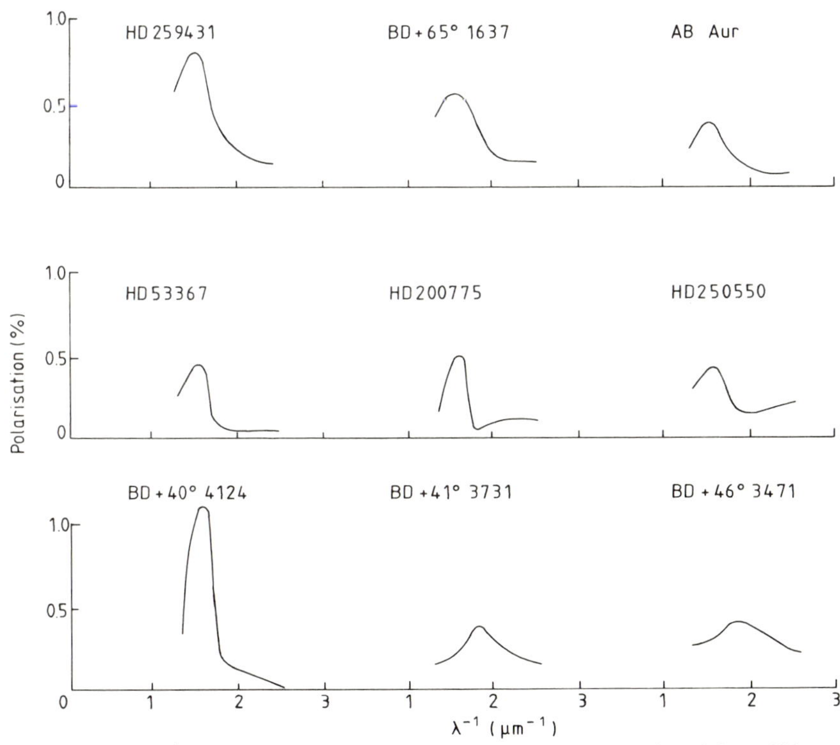

Figure 5.36. Polarisation curves of Herbig Ae/Be stars. (Reproduced from Vrba 1975 by permission.)

650 nm (Vrba 1975). The mechanism producing the polarisation is therefore not electron scattering modified by hydrogen absorption, as for the normal Be stars, and may be due to the dust surrounding the stars. The polarisation is variable in the majority of the stars (Breger 1974, Vrba *et al* 1979).

The gross spectroscopic and photometric properties of these stars show many similarities with those of the T Tauri stars. Finally, most of them have surface gravities less than or equal to zero-age main sequence stars. The interpretation of them as very young objects therefore seems highly likely to be correct.

It is possible that even hotter T Tauri analogues may exist amongst the O type stars, but may be masked by thick circumstellar dust (Strom *et al* 1972). Possible candidates could be η Car (see the next chapter) and V 645 Cyg where a possible O9 type star is observed as an Ae star due to masking by dust (Harvey and Lada 1980).

6. Other Emission Line Systems

6.1. Close Binary Systems

These are an important group of the emission-line stars, but many of their properties have already been covered earlier since they form subgroups of all the systems looked at so far. The emission lines almost always originate in the material flowing between the stars, which is due to one component filling its Roche lobe and spilling over on to the other star through the inner Lagrangian point.

Close binaries are usually classified into detached (algol type), semidetached (ellipsoidal, or β Lyra type when eclipsing) and contact (W UMa type) systems (figure 6.1). Significant mass exchange, or mass loss, leading to the production of emission lines, occurs in the latter two types where at least one star fills its Roche

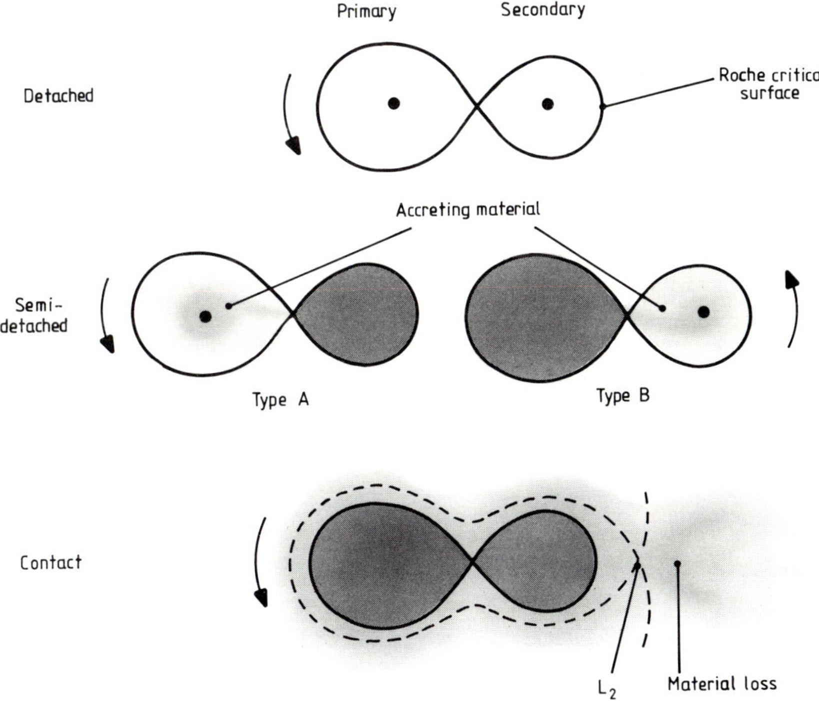

Figure 6.1. Types of binary system.

critical surface and is losing material. (Note that Algol, the star, is actually a semidetached system of type A! The confusion arose because the more massive and brighter star of the system is small compared with its critical Roche surface, and most observations of Algol relate to this star. The secondary may only be observed with difficulty. The emission lines from Algol discussed later in this chapter should not therefore be taken to imply the presence of emission lines in algol type stars.)

The behaviour of a close binary is largely governed by the Roche surfaces – surfaces of equal potential in the co-rotating frame of reference – and the sizes of the stars in comparison. The critical surface (shown by the bold curve in figure 6.2) defines two regions (the Roche lobes) wherein material is bound to one or other star. Outside the critical surface the material (which is assumed to be co-rotating with the binary) remains bound to the system as a whole unless it gets beyond the surface through L_3 when it may be lost entirely. The points marked L_1 to L_5 are called the Lagrangian points, and are positions of equilibrium in the co-rotating frame of reference. In most binaries they are positions of unstable equilibrium, but for small masses L_4 and L_5 can become points of stable equilibrium when

$$\frac{M_2}{M_1+M_2}<0.0385 \qquad (6.1)$$

(resulting in the Trojan asteroids in the Sun–Jupiter binary system). These surfaces depend only on gravitational and inertial forces. If other forces (e.g. radiation pressure) are significant, or if the material does not co-rotate with the binary, then they become very distorted (figure 6.3) and a contact system may no longer be possible. The average radius of a Roche lobe is given by equations (4.6) and (4.7), and its variation with mass ratio, together with the change in the positions of the centre of mass and the inner Lagrangian point, is shown in figure 6.4. Adequate models do not exist to fully describe the flow of material between the stars taking into account convection, turbulence, Coriolis forces etc. Convection, at least is likely to be important in a radiation pressure dominated accretion disc (Tayler 1980). A simplified model which fits the energy transfer in a close binary has been described by Robertson (1980).

For those binaries which have not already been mentioned in previous chapters, the emission arises from the stream of gas between the two stars and/or from a gaseous envelope surrounding the whole system. The latter may be co-rotating or it may be detached. The amount of material in the envelope or gas stream is small and difficult to observe, but may be detectable for favourable orientations of the orbit.

In the semidetached systems of type A, faint emission lines may sometimes be observed immediately after the start, and immediately before the end, of primary eclipse. These arise in the accretion disc of material around the primary (figure 6.5). In U Cephei there is a faint emission component to Hα at phase 0.79 which is not present at phase 0.52 (Batten 1973) (figure 6.6). The stream of gas between the two stars in Algol may be observed by its hydrogen emission lines which become

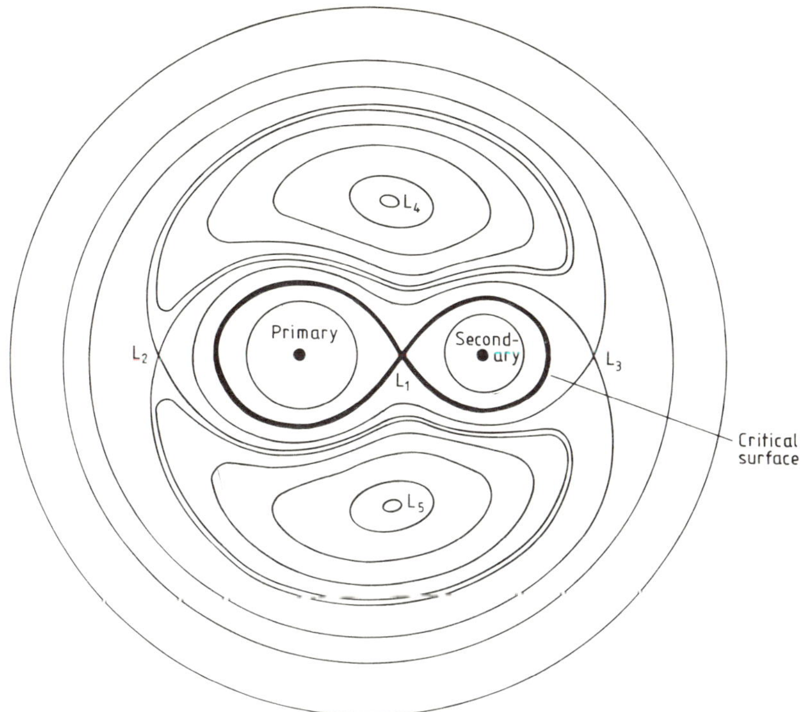

Figure 6.2. Roche equipotential surfaces for a mass ratio of 0.6. (Reproduced from Kopal 1978 by permission.)

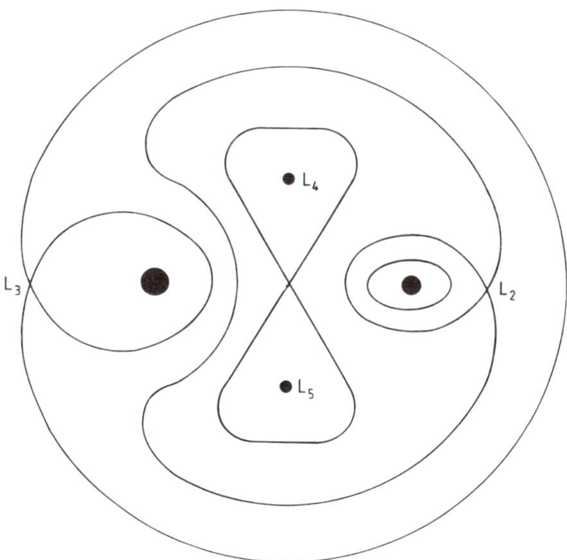

Figure 6.3. Roche equipotential surfaces including significant radiation pressure. (Equal masses for the components, radiation pressure 80% of the gravitational force for the left-hand star, and 50% of the gravitational force for the right-hand star.) (Reproduced from Sahade and Wood 1978 by permission.)

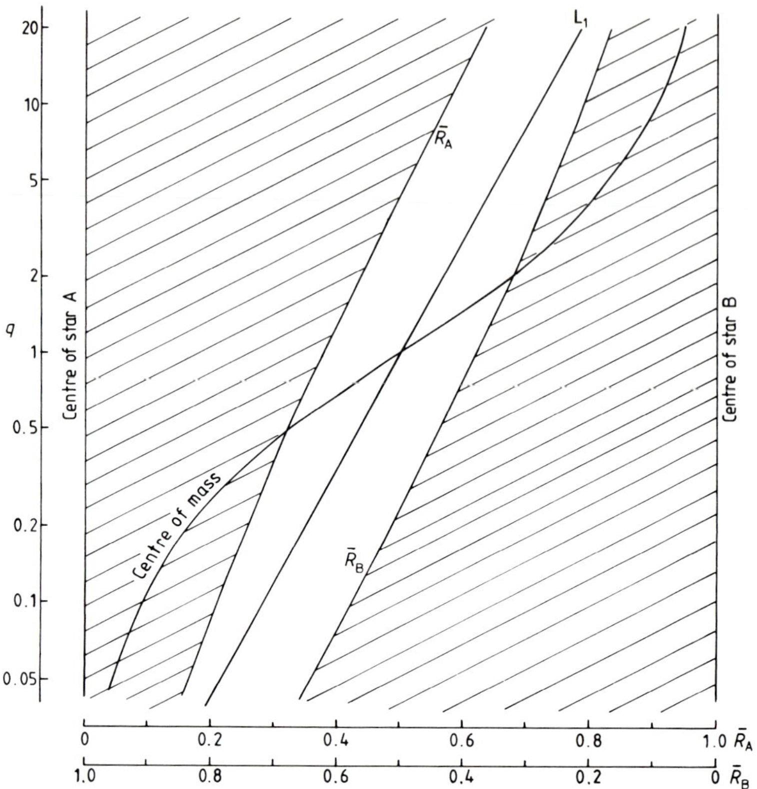

Figure 6.4. Variation of the mean size of the Roche lobes and the positions of the centre of mass and inner Lagrangian point with mass ratio ($q = M_A/M_B$). Units of distance are in terms of the separation of the stars.

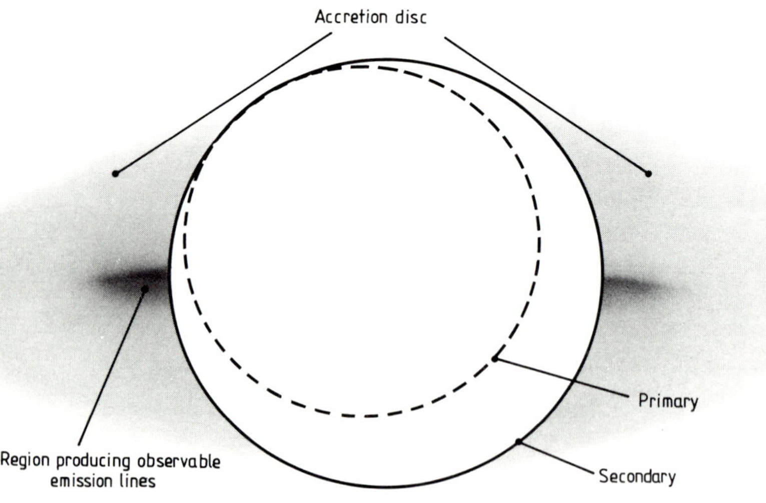

Figure 6.5. Schematic view of a type A semidetached system at primary eclipse.

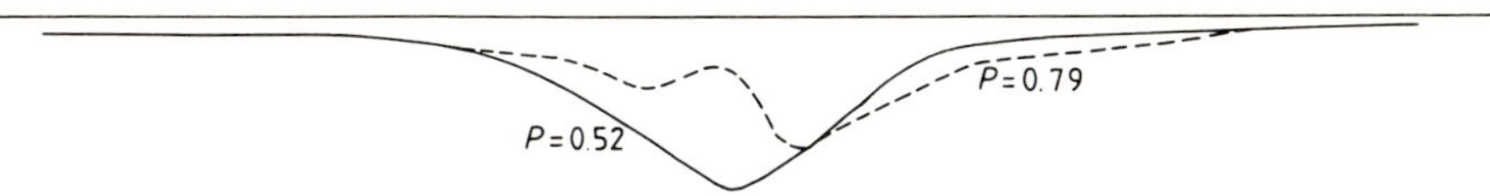

Figure 6.6. Profiles of Hα in U Cep at phases 0.52 and 0.79 from primary eclipse. (Reproduced from Batten 1973 by permission.)

visible at quadrature, and by emission cores to its Mg II 280 lines at the same time (Chen and Wood 1976). Contact binaries usually show Ca II H and K emission, and the region emitting the lines seems to be associated with the more massive component (Struve 1950).

The velocity of the material in the accretion stream and disc may distort the velocity curve of the binary. Sometimes the effect is only noticeable at particular phases, as in U Cep (figure 6.7). In other stars, the whole velocity curve is distorted and becomes very difficult to interpret (figure 6.8).

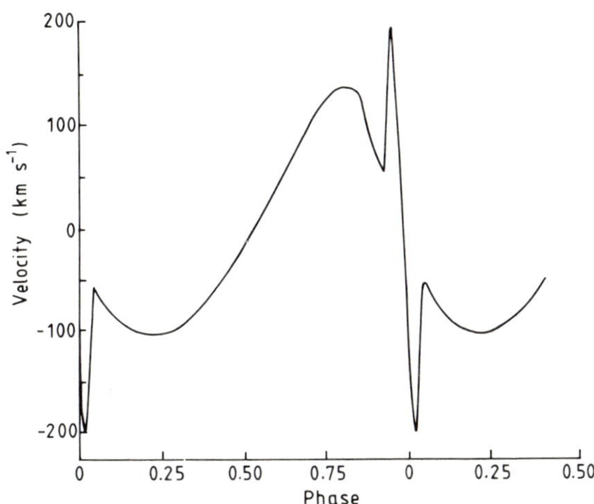

Figure 6.7. Velocity curve of U Cep. (Reproduced from Struve 1944 by permission.)

Many of the binaries dealt with earlier emit x-rays, sometimes very strongly. The x-ray emission usually arises from the impact of the accretion stream on the accretion disc. Less energetic emissions occur in systems like Algol which has variable 0.5 to 4.5 keV emission, and whose 2 to 6 keV emission reaches 3.8×10^{-14} W m^{-2} keV^{-1} (Harnden *et al* 1977, Schnopper *et al* 1976, White *et al* 1980b).

Centrimetric radio emission occurs in some binaries. In Algol the emission is erratic with strong outbursts lasting 16 hours or so, interspersed by long quiescent intervals. The emission seems likely to be associated with the mass exchange process, possibly also involving magnetic fields (Soderhjelm 1980), and with

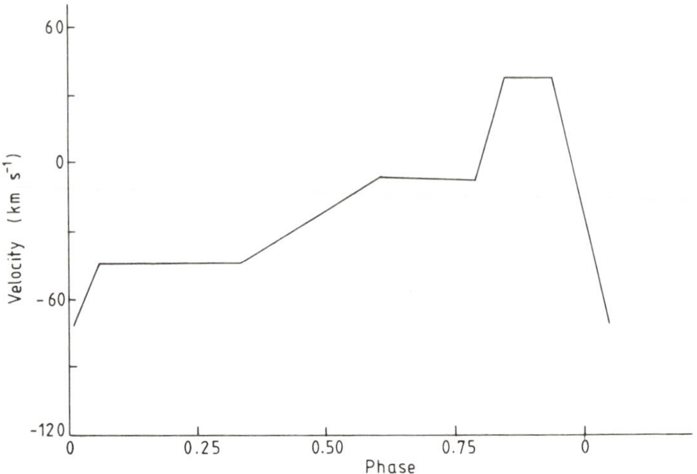

Figure 6.8. Velocity curve of RZ Scu. (Reproduced from Batten 1973 by permission.)

energetic electrons powered by the infalling matter producing incoherent synchrotron radiation (Hjellming *et al* 1972, Woodsworth and Hughes 1976). The RS CVn type binary systems, although not dealt with any further in this work being usually of spectral types G and K, are worth mentioning in passing as having the general property of being radio emitters. Their radio emission shows an erratic flare pattern similar to the emission from Algol, and is frequently circularly polarised.

β Lyrae has a secondary which is several times more massive than its primary, but which is too faint to be detectable in the visible. Emission lines occur, particularly at Hα and in the ultraviolet. The emission line variations show only weak, or no correlation with the binary phase. The system has an infrared excess which may originate as free-free emission in an optically thin region (Phillips *et al* 1980). There is also an ultraviolet excess whose origin is not entirely clear. Some of the ultraviolet lines have P Cygni profiles, whose corresponding expansion velocities are around 150 km s^{-1}. The mass loss is estimated at $10^{-5} M_\odot$ a^{-1} (Ziolkowski 1976), and the period of the binary is increasing as a result. As mentioned earlier the system is also a radio source. A possible model is shown in figure 6.9, based on the results of the intensive observing campaign conducted on the star in 1971 (Flora and Hack 1975).

Probably the most remarkable object discovered in the last decade is the system SS433. It is a 14th magnitude object in Aquila which was catalogued by Stephenson and Sanduleak (1977) as an emission line star. In 1974 it was identified with a small variable radio source (Clark and Crawford 1974, Clark *et al* 1975, Ryle *et al* 1978, Seaquist *et al* 1979) (figure 6.10) which was central to the large extended radio source, W50. Then in 1978, it was identified with the x-ray source A 1909 + 04 (Clark and Murdin 1978), with additional diffuse emission from two linear radial regions about 20″ long on either side (Seward *et al* 1980), and shown to be at

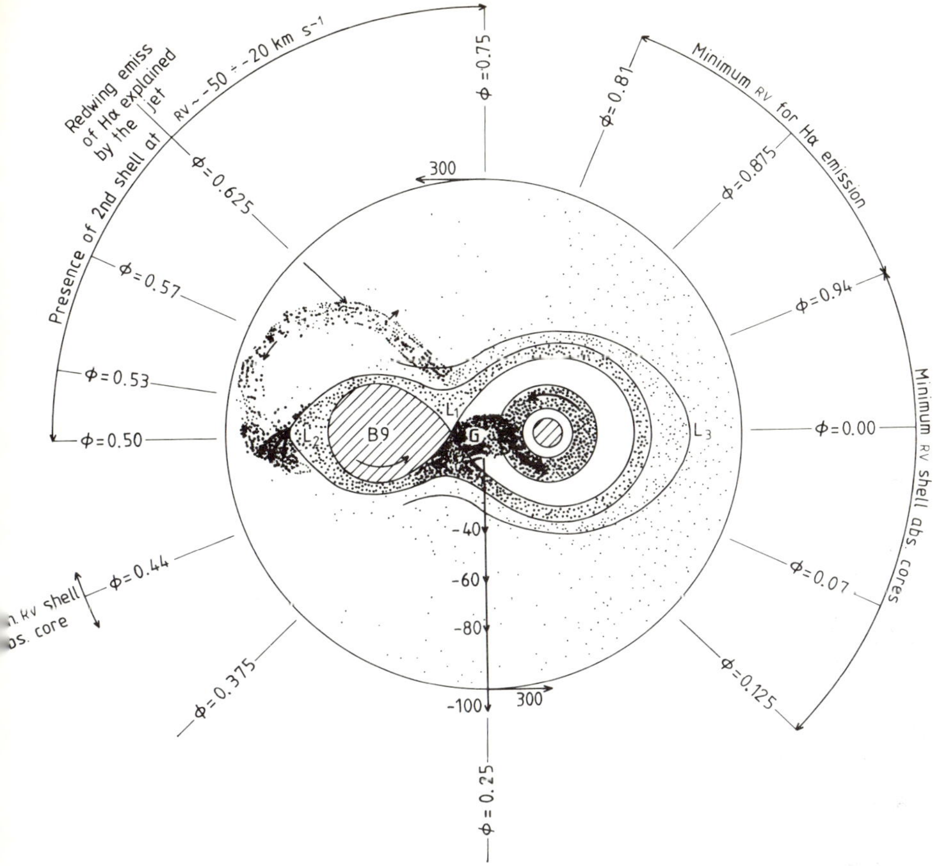

Figure 6.9. Schematic model of β Lyr. (Reproduced from Flora and Hack 1975 by permission.)

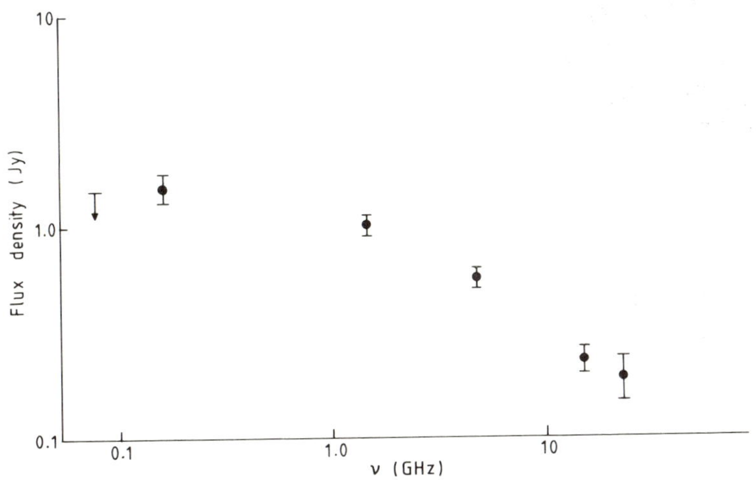

Figure 6.10. Radio spectrum of SS433. (Reproduced from Seaquist *et al* 1980 by permission.)

approximately the same distance as W50. These indications of peculiarity led in 1978 and 1979 to the study of the system spectroscopically (Clark and Murdin 1978, Liebert *et al* 1979, Margon *et al* 1979b). The really unusual nature of the system then became apparent. The spectrum was mainly comprised of very strong emission lines of neutral hydogen and helium. But most of these lines had three primary components – one nearly undisplaced, the other two displaced by up to 35 000 km s^{-1} to longer and shorter wavelengths (figure 6.11). Furthermore the displaced components oscillated about displacements of -6000 km s^{-1} for the (normally) blue-shifted component, and $+25\,000$ km s^{-1} for the (normally) red-shifted component with a period of 164.0 ± 0.3 days (Margon *et al* 1979a, 1980) (figures 6.12, 6.13, 6.14). In the infrared there is a continuum excess which corresponds to a black body at 15 700 K, but which may be due to free–free emission (Kundt 1979). Superimposed upon this are very strong Paschen and Brackett emission lines (Giles *et al* 1979, McAlary and McLaren 1980). The infrared flux may have an 11.8 day variation (Giles *et al* 1980) but this has not been found at other wavelengths. A 13 day period has been found for the 'fixed' lines (Crampton *et al* 1980) and for the optical flux (Kemp *et al* 1980). The absolute magnitude is estimated at -7^{m}, with A_{v} about 8^{m} and the distance about 4 kpc. The optical temperature is about 25 000 K, and the surface gravity is comparable with that of an early supergiant (i.e. about 10 m s^{-2}). The fixed lines correspond to a spectral type of O5f (Murdin *et al* 1980).

The object is so recently discovered that no consensus on a model to fit the data has yet been reached. In fact some 384 possibilities are currently suggested! (*New Scientist* 1980). The Martin–Rees model is summarised below in the belief that it is likely to be the survivor (or very similar to the finally agreed model) of these

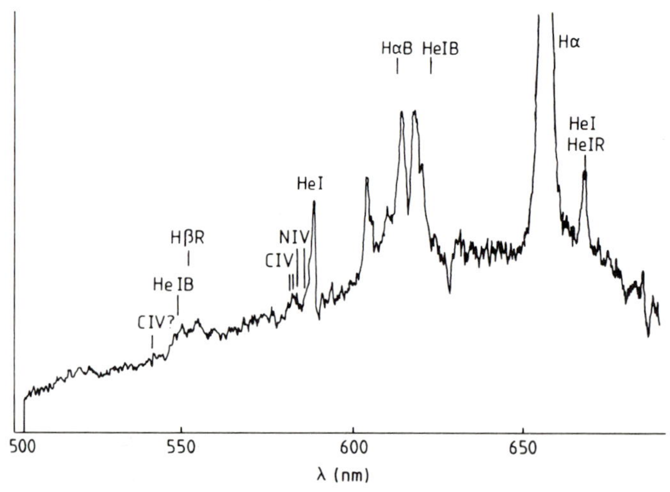

Figure 6.11. Spectrum of SS433 on 29 June 1978, showing Hα and its blue-shifted component, He I 587.6 (all components) and He I 667.8 and its blue-shifted component. (Reproduced from Murdin *et al* 1980 by permission.)

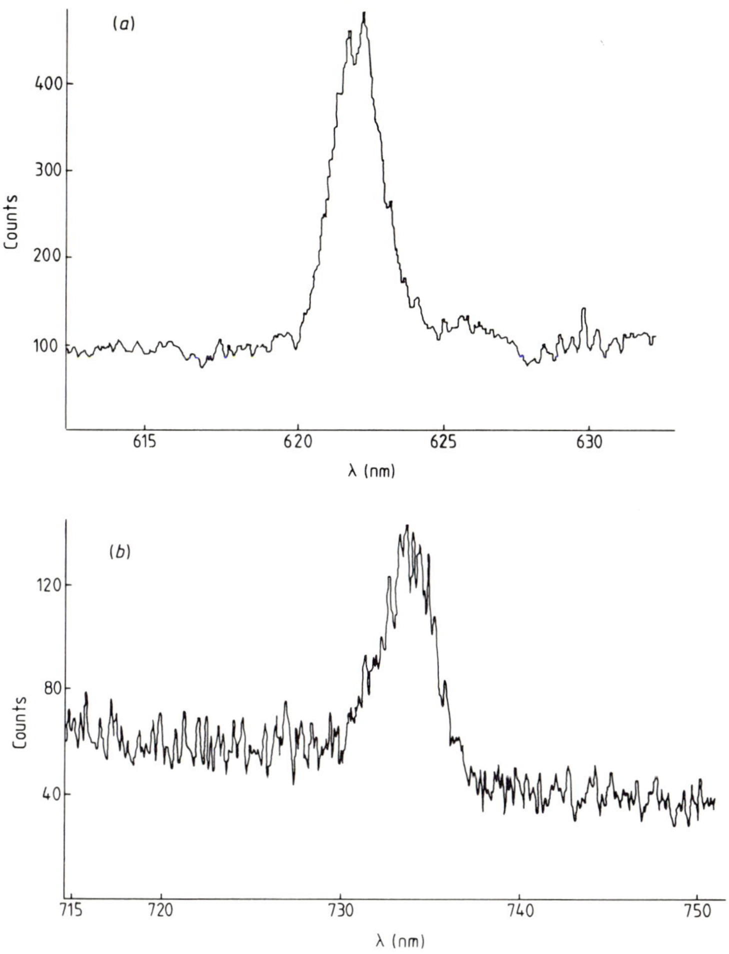

Figure 6.12. (*a*) Blue and (*b*) red moving components of Hα in SS433 on 13 September 1979. (Reproduced from Murdin *et al* 1980 by permission.)

numerous theories. References to some of the other models may be found in Abell and Margon (1979), Amitai-Milchgrub *et al* (1979a, b), Amitai-Milchgrub and Shaham (1979), Bedoghi *et al* (1980), Begelman *et al* (1980), Bonometto and Caderni (1980), Collins and Newson (1979), Fabian (1980), Fabian and Rees (1980), Fang (1981), van den Heuvel *et al* (1980), Kundt (1979), Liebert *et al* (1979), Milgrom (1979a, b, 1980), Seward *et al* (1980), Shaham (1980), Terlevich and Pringle (1979), Whitmire and Mateze (1980, 1981), Young and Burbidge (1979). In the Martin–Rees model (Martin and Rees 1979) the system is a very close binary (separation 8.6×10^7 m) with a 0.4 solar mass white dwarf orbiting a 5 solar mass rotating black hole. Mass transfer occurs from the white dwarf to an accretion disc around the black hole. Some material falls into the black hole, the rest is expelled at $0.25c$ along the rotational axis of the black hole (cf figure 1.6).

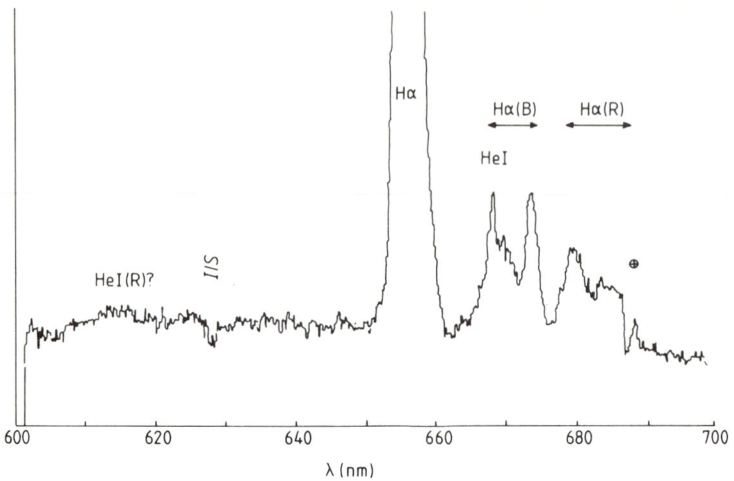

Figure 6.13. Spectrum of SS433 on 30 June 1979 showing the three components of Hα. (Reproduced from Murdin *et al* 1980 by permission.)

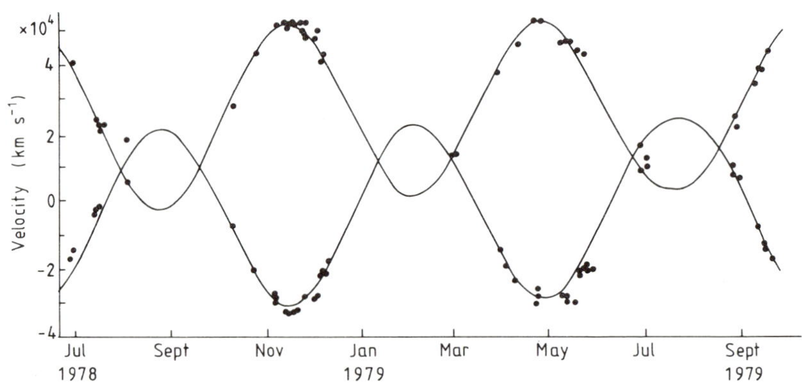

Figure 6.14. Velocities of the moving components of Hα in SS433. (Reproduced from Murdin *et al* 1980 by permission.)

If the black hole is rotating rapidly, it will drag the inertial frames of reference around with it (Lense-Thirring effect). If this rotational axis is inclined to that of the orbital motion and there is an accretion disc of 10^{-5} to $10^{-2}\,M_\odot$, then the same effect will cause the rotational axis to precess around the perpendicular to the orbital plane (Sarazin *et al* 1980). The jets along the rotational axis will therefore swing around with the same period as the precession. In SS433 the rotational axis is inclined by 17° to the perpendicular to the orbital plane, which in turn is inclined at 78° to the line of sight (these angles may be swapped around, since there is a two-fold ambiguity in the model). The model is shown schematically in figure 6.15. The precession period is thus 164 days (Margon and Abell 1979), and the moving emission components in the spectrum are formed in the jets. Their wavelengths change as precession of the black hole's rotational axis changes the inclination of

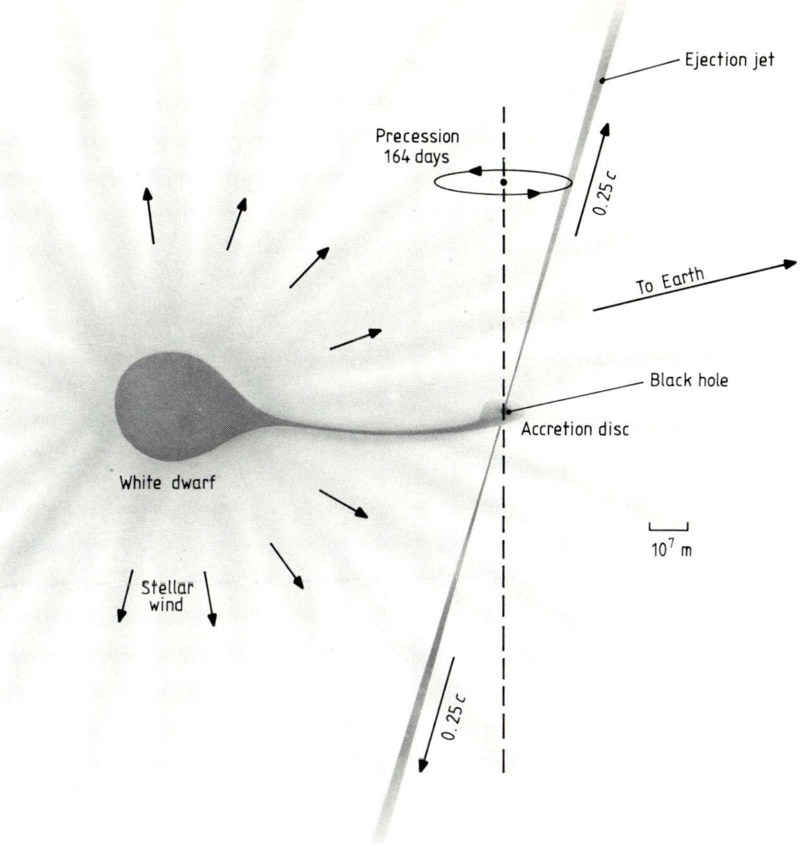

Figure 6.15. Schematic view of the Martin–Rees model of SS433.

the jets to the line of sight. The sub-components which may be found within the main components (figures 6.11 and 6.13) are due to pulses within the jets. Since the two moving components tend to be mirror images of each other (Murdin *et al* 1980), the (unknown) process producing these pulses must be common to both jets. The x-ray emission also probably arises in the jets. Most of the material from the white dwarf, however, is lost as a stellar wind, and not in the jets. This material forms a dense cloud around the binary and produces the continuum emission and the static emission lines. The width of the static emission lines suggests a wind velocity of 2000 km s^{-1}, and a mass loss rate of $3 \times 10^{-6} M_\odot a^{-1}$ (Zealey *et al* 1980). The radius of the region of unit optical depth for electron scattering is about 3×10^9 m. The binary period is unlikely therefore to be detectable. (The orbital period is likely to be only a few minutes, so that neither the 11.8 nor 13 day periods may be attributed to it. These are more likely to arise from motion of the hot spot on the accretion disc or from a third member of the system (Begelman *et al* 1980, Margon *et al* 1980).) The velocity of the moving components when they are superimposed is about +10 000 km s^{-1}. Since this represents the instant when the direction of motion of the jets is perpendicular to the line of sight, the red-shift is attributable to the transverse Doppler effect (Milgrom 1979a). The associated radio source, W50, was at one time regarded as a supernova remnant, but if the jets of SS433 have existed for 10 000 years or thereabouts, then they possess sufficient kinetic energy to energise and 'inflate' this source (Martin and Rees 1979). There is some evidence for two nebulosities in W50 which are on opposite sides of SS433 and are separated from it by about 70 arc minutes (Zealey *et al* 1980). These may be due to the interactions of the jets with the interstellar medium.

SS433 is unique at the moment. However, two other systems, CTB/G109.1 – 1.0 and HD 44179 ('the Red Rectangle'), exhibit some, but not all, of its peculiarities.

6.2. Supergiants

Early supergiants usually exhibit a faint emission component to Hα which forms a P Cygni type of line profile. Emission may also occur at Hβ, He I 468.6, 587.6, 667.8, C III 569.6 and N III 463.4–464.1. The emission is usually variable and may sometimes disappear completely (figures 6.16 and 6.17). The optical line profiles suggest expansion velocities of 100 km s^{-1} or thereabouts. However, in the ultra-violet, resonance lines of high ions of silicon, carbon, nitrogen and oxygen suggest

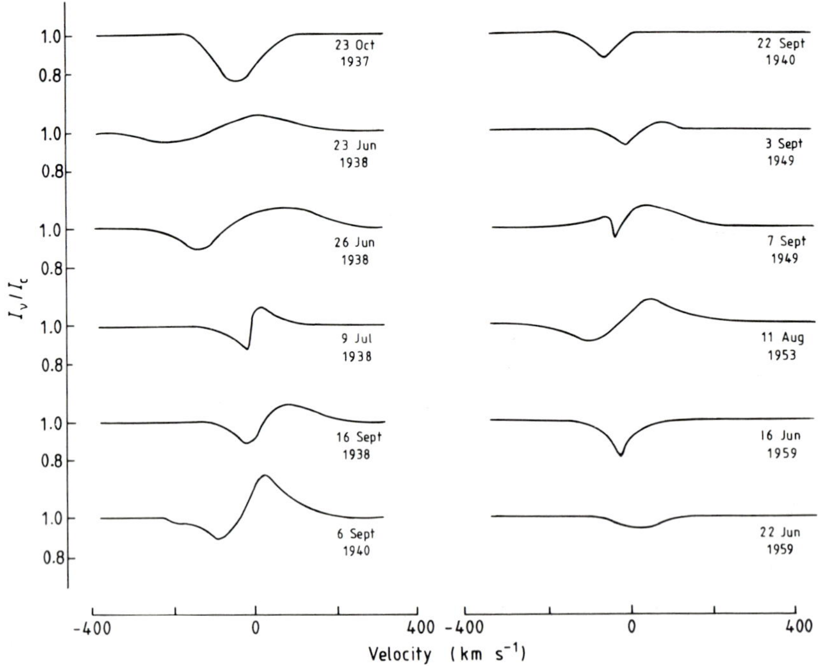

Figure 6.16. Hα profiles in the B3Ia star, 55 Cyg. (Reproduced from Underhill 1966 by permission.)

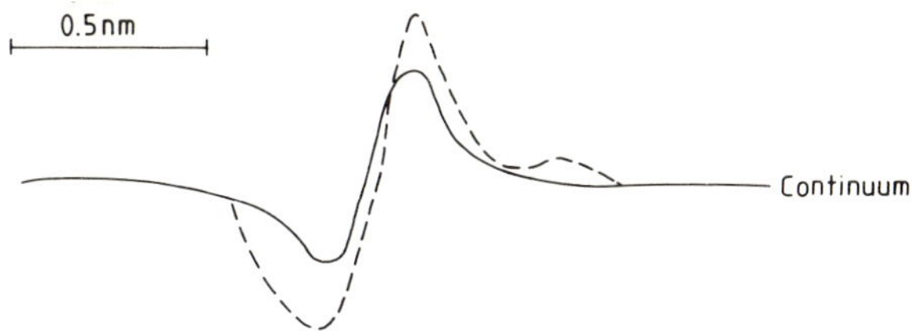

Figure 6.17. He I 587.6 in the B1 Ia–O star, HD 152236 in 1973. The two profiles are from spectra separated by 30 days. (Reproduced from Wolf and Sterken 1979 by permission.)

velocities of thousands of kilometres per second (Morton 1967) leading to mass loss rates of about $10^{-6} M_\odot$ a^{-1}. (The very high levels of ionisation, N V, O VI, etc, may originate through Auger ionisation due to K-shell absorption of x-rays.) ζ' Sco, an extreme B1/2 Ia supergiant has a mass loss rate which averages $1.5 \times 10^{-5} M_\odot$ a^{-1}. It is not a steady mass loss, however, but occurs as a series of 'puffs' leading to multiple absorption components in the Mg II 280 lines (Heck *et al* 1980) (figure 6.18). The velocity of these 'puffs' ranges from 180 to 400 km s^{-1}. The apparent visual magnitude also varies by up to 0.1^{m}, in a manner which is loosely correlated with increased activity in the intermittent mass loss process.

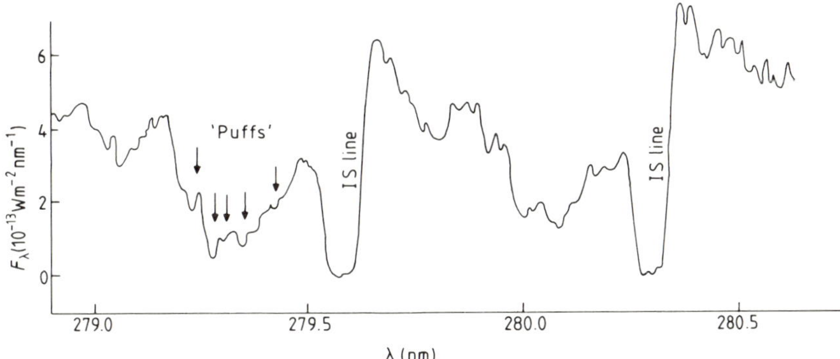

Figure 6.18. Mg II 280 lines in ζ' Sco. Obtained on 21 June 1979. The absorption components due to several 'puffs' are indicated by arrows for the 279.5 line. The deep absorptions are due to the interstellar gas. (Reproduced from Heck *et al* 1980 by permission.)

The absolute visual magnitudes range from -7.5^{m} to -8.5^{m} for the O type supergiants through -9.0^{m} for the B type to -9.5^{m} for the F type supergiants (Humphreys 1978). Most vary by up to 0.06^{m} with timescales of 5 to 100 days (Maeder and Rufener 1974). The cause of the variations seems likely to be radial pulsations of the star, and their irregular velocity variations of up to ± 30 km s^{-1} (Underhill 1966) may be associated with such pulsations. The velocity variations may also provide the mechanism which initiates the high-velocity stellar wind observed in the ultraviolet, since they exceed the velocity of sound.

The VV Cephei binaries are massive long-period systems. One component is a cool supergiant and the other an OB star. Balmer emission lines are found from an envelope around the hot star. Emission lines of [Fe II] are also observed indicating the presence of very low-density gas near the system (Mollenhoff and Schaifers 1978).

6.3. P Cygni and Related Stars

P Cygni is a well known star because it lends its name to the very common P Cygni type of line profile. This has an emission component which is usually undisplaced, and an absorption component displaced to shorter wavelengths (figure 6.19). Many of the stars discussed in this book have spectra which contain P Cygni line profiles, especially in the ultraviolet. The presence of such a line profile is usually indicative of outward motions in the star's atmosphere.

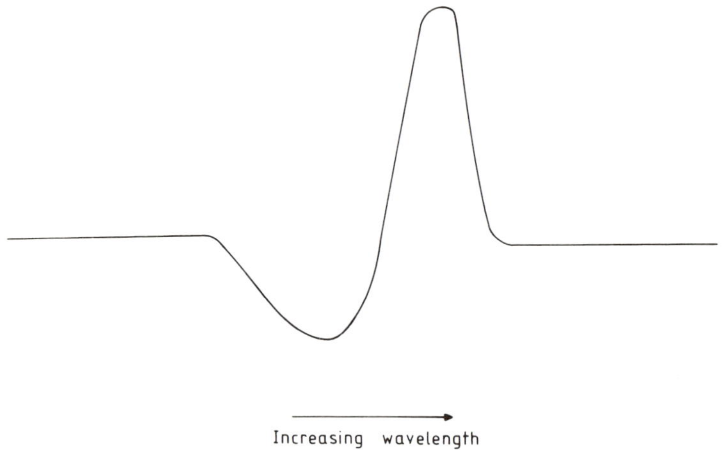

Figure 6.19. A P Cygni type of line profile.

P Cygni itself, however, is something of an enigma. Various workers have classed it with novae, shell stars, Hubble-Sandage variables, Wolf-Rayet stars, Be stars, close binary systems, post supergiant stars, supernovae, W UMa systems and planetary nebulae! The history of its observation extends over at least four centuries. It first called attention to itself by a sudden increase in brightness in the year 1600. Thereafter it varied markedly until 1715, since when it has been more or less constant (figure 6.20). At its peak, P Cyg reached an absolute bolometric magnitude of -11^{m}.

The spectrum of P Cyg is that of a B1.5 supergiant with additional emission lines (its continuum is shown in figure 6.21). The hydrogen and helium emission lines lead to estimates of 10 000 to 15 000 K for the electron temperature, and 25 000 K for the radiation temperature (Underhill 1966). The ultraviolet continuum suggests a temperature of 12 200 K (Underhill 1979). The displacements of the absorption

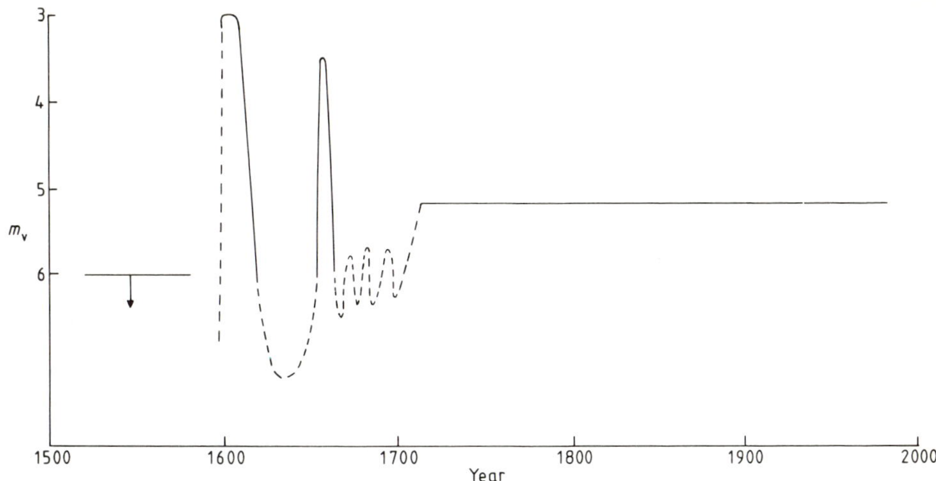

Figure 6.20. Schematic light curve of P Cyg since 1500 AD (Beals 1950).

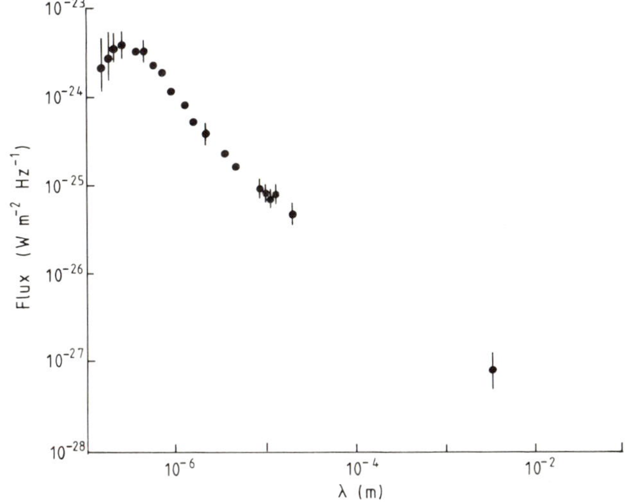

Figure 6.21. Continuous spectrum of P Cyg. (Reproduced from Nugis *et al* 1979 by permission.)

components suggest expansion velocities ranging from 30 to 230 km s^{-1}, with there being a tendency for those lines with the larger expansion velocities to be produced from the lower excitation and ionisation states. In the ultraviolet, velocities of up to 500 km s^{-1} are found from the resonance lines of carbon, nitrogen and silicon (Cassatella *et al* 1979, de Groot 1973, Hutchings 1979b), but the C III 117.5 line shows no velocity shift at all (Hutchings 1976a). It therefore seems likely that the atmosphere is accelerating outwards, since the coolest parts of the shell are to be expected furthest from the star. (A range of line profiles for a variety of outwardly accelerating flows in contact and detached shells have been calculated by Castor and Lamers (1979), and these do match the profiles in P Cyg, but Kuan and Kuhi (1975) find that a deccelerating flow is preferable.)

The rate of mass loss is estimated between 2×10^{-5} and $3.5 \times 10^{-4} M_\odot\ a^{-1}$ by several methods (Abbott *et al* 1980, van Blerkom 1978a, de Groot 1969, Nugis *et al* 1979). The star may have been detected at 90 GHz (intensity 82 ± 40 mJy), and a mass loss rate of $2 \times 10^{-5} M_\odot\ a^{-1}$ is thereby inferred (Schwartz and Spencer 1977). A very much lower rate of mass loss of only 1.5 to $8 \times 10^{-6} M_\odot\ a^{-1}$ has been found by Underhill from an estimate of the 3 GHz emission obtained by extrapolation from the infrared excess (Underhill 1979). A similarly low rate of mass loss has been obtained from a model of the He I emission using a 36-level helium atom (Oegerle and van Blerkom 1976). The lack of change in the light curve suggests that mass losses at a rate of perhaps $5 \times 10^{-5} M_\odot\ a^{-1}$ have been continuing since 1715 – a total mass loss of perhaps $0.01 M_\odot$. Estimates of dilution lead to shell sizes of two to three stellar radii, but the shell must obviously be very much larger. An expansion at even 200 km s^{-1} for 250 years leads to a radius of 0.05 pc.

P Cyg does not form the prototype of an easily identifiable and homogeneous class of stars. HD 94910 (AG Car), HD 37836 and HD 269858 are closest to P Cyg in their appearance. Beals (1950) lists 69 P Cygni type stars, but these include supergiants, Of stars, β Lyr stars and a planetary nebula. Eliminating these, only about half may be considered 'true' P Cygni type stars in the sense that their spectra contain a number of P Cygni line profiles for several elements in the optical region. Of this group P Cyg is the most extreme member, being of earlier spectral type than most, and having the characteristics of the group most highly developed. Their general features are therefore similar to those already outlined for P Cyg itself (de Groot 1973):

(*a*) an early type absorption spectrum,

(*b*) P Cygni line profiles for lines of H, He I, N I, N II, N III, Fe II, Fe III, Ti II, etc,

(*c*) expansion velocities from tens to thousands of kilometres per second with the highest velocities associated with the lowest excitation/ionisation lines,

(*d*) irregular, low-amplitude light variations (only P Cyg and AG Car (Mayall 1969) have varied by more than two magnitudes),

(*e*) mass loss rates of 10^{-5} to $10^{-4} M_\odot\ a^{-1}$.

Their absolute magnitudes are variously estimated from -3^m to -8.5^m (Arkhipova 1964, Luud 1967).

AG Car is surrounded by a low-excitation nebula a third of a parsec across with an estimated electron density of $1.7 \times 10^9\ m^{-3}$, an electron temperature of 7500 to 10 000 K, and mass of 0.18 to $0.25 M_\odot$ (Johnson 1976, Thackeray 1977b). Its spectrum possesses 'twisted' spectral lines indicating that it is in a highly turbulent state. It may alternatively, be interpreted as an extreme planetary nebula.

The star, η Car has been suggested as an even more extreme form of this group of stars (also as a slow nova and a B[e] star). It brightened to an apparent visual magnitude of minus one in 1840, then faded to around 7^m or 8^m, rising in the 1950s to its current brightness of about $+6.0^m$. At its peak it reached an absolute bolometric magnitude of -13^m. About 99% of its emission is in the infrared, so that its M_{bol} is still around -12^m (Andriesse and Viotti 1979). Its mass loss rate is estimated at 0.075 $M_\odot$ a^{-1} (Andriesse and Viotti 1979), and its mass at 160 $M_\odot$. The infrared emission arises from dust which is estimated to be forming at a rate of $1.5 \times 10^{-4} M_\odot$ a^{-1} (Andriesse *et al* 1978) (or, less plausibly, due to infrared emission lines of Fe II and Fe III (Thackeray 1978)). There is a feature at 10 μm which is probably due to silicates in the dust. About 10^{-7} of the system's energy is emitted as x-rays over the area of the shell of η Car, with a bright spot which may be the central object (Seward *et al* 1979).

Another candidate for an extreme P Cygni system is MWC 349 (Hartmann *et al* 1980) (figure 5.26). Other possibly related systems could include the slow novae (e.g. RR Tel and RR Pic) (chapter 4), Be stars (chapter 5) and symbiotic stars (chapter 4 – Z And became a P Cygni star in 1933, returning to its prior state of a combination spectrum of spectral types M and W some time later (Sobolev 1960)). The present state of the observational study of P Cygni stars is insufficient, however, to substantiate or eliminate any such relationships.

Bibliography

References for each detailed point are given within the text. This section lists a few reviews/books/symposia etc for each chapter which the reader may care to consult for further general information on the topic.

Chapter 1. Extended Atmospheres and the Origin of Emission Lines

Hummer D G 1976 *IAU Symp. No 70, Be and Shell Stars* ed A Slettebak (Dordrecht: Reidel) p281

Hummer D G and Rybicki G B 1967 *Methods in Computational Physics* 7 53

—— 1971 *Ann. Rev. Astron. Astrophys.* **9** 237

Mihalas D 1978 *Stellar Atmospheres* (San Francisco: W H Freeman)

National Bureau of Standards 1970 *Spectrum Line Formation in Stars with Steady State Extended Atmospheres* Special Publication No 332

Sobolev V V 1960 *Moving Envelopes of Stars* (Harvard University Press)

Chapter 2. Wolf-Rayet Stars

Bappu M K V and Sahade J (ed) *IAU Symp. No 49, Wolf-Rayet and High Temperature Stars* (Dordrecht: Reidel)

Conti P S 1978 *Ann. Rev. Astron. Astrophys.* **16** 371

Conti P S and de Loore C W H (ed) 1979 *IAU Symp. No 83, Mass Loss and Evolution of O Type Stars* (Dordrecht: Reidel)

National Bureau of Standards 1968 *Wolf-Rayet Stars* Special Publication No 307

Chapter 3. Planetary Nebulae

Gurzadyan G A 1970 *Planetary Nebulae* (Dordrecht: Reidel)

Miller J S 1974 *Ann. Rev. Astron. Astrophys.* **12** 331

Osterbrock D E 1974 *Astrophysics of Gaseous Nebulae* (San Francisco: W H Freeman)

Osterbrock D E and O'Dell C R (ed) 1968 *IAU Symp. No 34, Planetary Nebulae* (Dordrecht: Reidel)

Terzian Y (ed) 1978 *IAU Symp. No 76, Planetary Nebulae – Observation and Theory* (Dordrecht: Reidel)

Chapter 4. Cataclysmic Variables

Freidjung M (ed) 1977 *Novae and Related Stars* (Dordrecht: Reidel)

Gallagher J S and Starrfield S 1978 *Ann. Rev. Astron. Astrophys.* **16** 171

Glasby J S 1970 *The Dwarf Novae* (London: Constable)
McLaughlin D B 1960 *Stars and Stellar Systems* ed J L Greenstein (University of Chicago Press) **6** 585
Payne-Gaposchkin C 1957 *The Galactic Novae* (Amsterdam: North Holland)
Robinson E 1976 *Ann. Rev. Astron. Astrophys.* **14** 119
Warner B 1975 *IAU Symp. No 73, Structure and Evolution of Close Binary Systems* ed P Eggleton, S Mitton and J Whelan (Dordrecht: Reidel) p85

Chapter 5. Be Stars

Slettebak A (ed) 1976 *IAU Symp. No 70, Be and Shell Stars* (Dordrecht: Reidel)
—— 1979 *Space Sci. Rev.* **23** 541

Chapter 6. Other Emission Line Systems

Batten A H 1973 *Binary and Multiple Systems of Stars* (Int. Series of Monographs on Natural Philosophy Vol 51) (Oxford: Pergamon)
Kopal Z 1978 *Dynamics of Close Binary Systems* (Astrophysics and Space Science Library Vol 68) (Dordrecht: Reidel)
Sahade J and Wood F B 1978 *Interacting Binary Stars* (Int. Series of Monographs on Natural Philosophy Vol 95) (Oxford: Pergamon)

References

Abbott D C 1978 *Astrophys. J.* **225** 893

Abbott D C, Bieging J H, Churchwell E and Cassinelli J P 1980 *Astrophys. J.* **238** 196

Abell G O 1955 *Publ. Astron. Soc. Pacific* **67** 258

Abell G O and Margon B 1979 *Nature* **279** 701

Abney W de W 1877 *Mon. Not. R. Astron. Soc.* **37** 278

Acker A 1980 *Astron. Astrophys.* **89** 33

Acker A and Marcout J 1977 *Astron. Astrophys. Suppl.* **30** 217

Aitken D K, Barlow M J, Roche P F and Spenser P M 1980 *Mon. Not. R. Astron. Soc.* **192** 679

Ali-Alper M 1979 *Mon. Not. R. Astron. Soc.* **189** 305

Allen C W 1973 *Astrophysical Quantities* (3rd edn) (London: Athlone)

Allen D A 1979a *Mon. Not. R. Astron. Soc.* **189** 221

—— 1979b *Nature* **281** 284

Allen D A, Beattie D H, Lee T J, Stewart J M and Williams P M 1978 *Mon. Not. R. Astron. Soc.* **182** 57P

Allen D A, Hyland A R and Caswell J L 1980 *Mon. Not. R. Astron. Soc.* **192** 505

Aller L H 1969a *Sky and Telescope* **37** 348

—— 1969b *Sky and Telescope* **38** 306

—— 1969c *Sky and Telescope* **39** 15

Altamore A, Baratta G B, Cassatella A, Grasdalen G L, Persi P and Viotti R 1980 *Astron. Astrophys.* **90** 290

Amitai-Milchgrub A, Piran T and Shaham J 1979a *Nature* **279** 505

—— 1979b *Nature* **280** 473

Amitai-Milchgrub A and Shaham J 1979 *Astron. Astrophys.* **77** L7

Andriesse C D 1980 *Mon. Not. R. Astron. Soc.* **192** 95

Andriesse C D, Donn B D and Viotti R 1978 *Mon. Not. R. Astron. Soc.* **185** 771

Andriesse C D and Viotti R 1979 *IAU Symp. No 83, Mass Loss and Evolution of O Type Stars* ed P S Conti and C W H de Loore (Dordrecht: Reidel) p47

Andrillat Y and Houziaux L 1969 *Mass Loss from Stars* ed M Hack (Dordrecht: Reidel) p281

Anglo-Australian Observatory Newsletter 1980a, no 14

Anglo-Australian Observatory Newsletter 1980b, no 16

Apruzese J P 1974 *Astrophys. J.* **188** 539

Arkhipova V P 1964 *Sov. Astron.* **7** 686

Arnold S, Berg R A and Duthie J G 1976 *Astrophys. J.* **206** 790

Bahng J D R 1976 *IAU Symp. No 70, Be and Shell Stars* ed A Slettebak (Dordrecht: Reidel) p41
Bailey J 1978 *Mon. Not. R. Astron. Soc.* **185** 73P
—— 1979 *Mon. Not. R. Astron. Soc.* **189** 41P
—— 1980 *Mon. Not. R. Astron. Soc.* **190** 119
Bailey J, Jones D H P, Parkes G E and Mason K O 1978 *Mon. Not. R. Astron. Soc.* **184** 73P
Bailey J and Ward M 1981 *Mon. Not. R. Astron. Soc.* **194** 17P
Barker T 1979 *Astrophys. J* **227** 863
—— 1980 *Astrophys. J.* **237** 482
Barlow M J, Blades J C and Hummer D G 1980 *Astrophys. J.* **241** L27
Barlow M J and Cohen M 1977 *Astrophys. J.* **213** 737
Barnes T G, Lambert D C and Potter A E 1974 *Astrophys. J.* **187** 73
Bateson F M 1978 *Mon. Not. R. Astron. Soc.* **184** 567
Bath G T 1973 *Nature Phys. Sci.* **246** 84
—— 1977a *Mon. Not. R. Astron. Soc.* **178** 203
—— 1977b *Novae and Related Stars* ed M Freidjung (Dordrecht: Reidel) p41
—— 1978 *Mon. Not. R. Astron. Soc.* **182** 35
Bath G T, Pringle J E and Whelan J A J 1980 *Mon. Not. R. Astron. Soc.* **190** 185
Bath G T and Shaviv G 1976 *Mon. Not. R. Astron. Soc.* **175** 305
—— 1978 *Mon. Not. R. Astron. Soc.* **183** 515
Batten A H 1973 *Binary and Multiple Systems of Stars* (Int. Series of Monographs on Natural Philosophy, Vol 51) (Oxford: Pergamon)
Beals C S 1929 *Mon. Not. R. Astron. Soc.* **90** 202
—— 1930 *Publ. Dominion Astrophys. Observatory* **4** 271
—— 1931 *Mon. Not. R. Astron. Soc.* **91** 966
—— 1934 *Publ. Dominion Astrophys. Observatory* **6** 95
—— 1938 *Trans. IAU* **6** 248
—— 1940 *J. R. Astron. Soc. Canada* **34** 169
—— 1950 *Publ. Dominion Astrophys. Observatory* **9** 1
Beckwith S, Persson S E and Gatley I 1978 *Astrophys. J.* **219** L33
Bedoghi R, Braccesi A, Marano B and Messina A 1980 *Astron. Astrophys.* **84** L4
Beeckmans F and Hubert-Delplace A M 1980 *Astron. Astrophys.* **86** 72
Begelman M C, Sarazin C L, Hatchett S P, McKee C F and Arons J 1980 *Astrophys. J.* **238** 722
van den Bergh S 1974 *Astron. Astrophys.* **32** 351
Bernat A P, Barnes T G, Schupler B R and Potter A E 1977 *Publ. Astron. Soc. Pacific* **89** 541
Bertaud C 1948 *Annls Astrophys.* **11** 1
Bertiau F C and McCarthy M F 1969 *Ricerche Astron. Specola Vaticana* **7** 523
Bianchini A 1980 *Mon. Not. R. Astron. Soc.* **192** 127
Bidelman W P 1976 *IAU Symp. No 70, Be and Shell Stars* ed A Slettebak (Dordrecht: Reidel) p457
Blades J C and Whittet D C B 1980 *Mon. Not. R. Astron. Soc.* **191** 701
van Blerkom D 1978a *Astrophys. J.* **221** 186
—— 1978b *Astrophys. J.* **225** 175

van Blerkom D and Arny T T 1972 *Mon. Not. R. Astron. Soc.* **156** 91
Bode M F and Evans A 1980 *Astron. Astrophys.* **89** 158
Boeshaar G O 1974 *Astrophys. J* **187** 283
Boeshaar G O and Bond H E 1977 *Astrophys. J.* **213** 421
Bohannan B and Conti P S 1976 *Astrophys. J.* **204** 797
Bohm K H 1968 *IAU Symp. No 34, Planetary Nebulae* ed C R Osterbrock and C R O'Dell (Dordrecht: Reidel) p297
Bond H E 1978 *Publ. Astron. Soc. Pacific* **90** 216
Bond H E, Liller W and Mannery E J 1978 *Astrophys. J.* **223** 252
Bonometto S and Caderni N 1980 *Mon. Not. R. Astron. Soc.* **191** 411
Bowen I S 1934 *Publ. Astron. Soc. Pacific* **46** 146
—— 1935 *Astrophys. J.* **81** 1
Boyarchuk A A 1969 *Non-periodic Phenomena in Variable Stars* ed L Detre (New York: Academic) p395
Bradt H V, Apparao K M V, Clark G W, Dower R, Doxey R, Hearn D R, Jernigan J G, Joss P C, Mayer W, McClintock J and Walter F 1977 *Nature* **269** 21
Breger M 1974 *Astrophys. J.* **188** 53
Breysacher J and Vogt N 1980 *Astron. Astrophys.* **87** 349
Brocklehurst M 1971 *Mon. Not. R. Astron. Soc.* **153** 471
Bruhweiler F C, Morgan T H and van der Hucht K A 1978 *Astrophys. J.* **225** L71
Brune W H, Mount G H and Feldman P D 1979 *Astrophys. J.* **227** 884
Burton W M, Evand R G, Patchett B and Wu C-C 1978 *Mon. Not. R. Astron. Soc.* **183** 605
Cahn J H and Wyatt S P 1976 *Astrophys. J.* **210** 508
Calvet N and Cohen M 1978 *Mon. Not. R. Astron. Soc.* **182** 687
Campbell L 1919 *Harvard Annals* **48** 39
Campolonghi F, Gilmozzi R, Guidoni U, Messi R, Natali G and Wells J 1980 *Astron. Astrophys.* **85** L4
Canto J, Elliott K H, Meaburn J and Theokas A C 1980 *Mon. Not. R. Astron. Soc.* **193** 911
Carpenter G F, Eyles C J, Skinner G R, Willmore A P and Wilson A M 1976 *Mon. Not. R. Astron. Soc.* **176** 397
Cassatella A, Beeckmans F, Benvenutti P, Clavel J, Heck A, Lamers H J G L M, Macchetto F, Penston M, Selvelli P C and Stickland D 1979 *Astron. Astrophys.* **79** 223
Castor J I 1970 *Mon. Not. R. Astron. Soc.* **149** 111
—— 1979 *IAU Symp. No 83, Mass Loss and Evolution of O Type Stars* ed P S Conti and C W H de Loore (Dordrecht: Reidel) p175
Castor J I, Abbott D C and Klein R I 1975 *Astrophys. J.* **195** 157
Castor J I and Lamers H J G L M 1979 *Astrophys. J. Suppl.* **39** 481
Chaisson E J and Malkan M A 1976 *Astrophys. J.* **210** 108
Chanan G A, Nelson J E and Margon B 1978 *Astrophys. J.* **226** 963
Chanmugen G and Wagner R L 1977 *Astrophys. J.* **213** L13
Chen K-Y and Wood F B 1976 *Mon. Not. R. Astron. Soc.* **176** 5P
Chiappetti L, Tanzi E G and Treves A 1980 *Space Sci. Rev.* **27** 3
Chu Y-H and Lasker B M 1980 *Publ. Astron. Soc. Pacific* **92** 730

Churchwell E, Terzian Y and Walmsley M 1976 *Astron. Astrophys.* **48** 331
Ciatti F, Mammano A and Vittone A 1977 *Astron. Astrophys.* **61** 459
—— 1978 *Astron. Astrophys.* **68** 251
—— 1979 *Astron. Astrophys.* **79** 247
Clark D H and Crawford D F 1974 *Australian J. Phys.* **27** 713
Clark D H, Green A J and Caswell J L 1975 *Australian J. Phys., Astrophys. Suppl.* **37** 75
Clark D H and Murdin P 1978 *Nature* **276** 45
Cohen M 1975 *Mon. Not. R. Astron. Soc.* **173** 489
—— 1980 *Mon. Not. R. Astron. Soc.* **191** 499
Cohen M, Barlow M J and Kuhi L V 1975 *Astron. Astrophys.* **40** 291
Cohen M, Hudson H S, O'Dell S L and Stein W A 1977 *Mon. Not. R. Astron. Soc.* **181** 233
Cohen M and Kuhi L V 1977 *Mon. Not. R. Astron. Soc.* **180** 37
Cohen M and Vogel S N 1978 *Mon. Not. R. Astron. Soc.* **185** 47
Collins G W II 1974 *Astrophys. J.* **191** 157
Collins G W II and Foltz C B 1977 *Publ. Astron. Soc. Pacific* **89** 596
Collins G W II and Newson G H 1979 *Nature* **280** 474
Collins G W II and Sonneborn G H 1977 *Astrophys. J. Suppl.* **34** 41
Colvin J D, van Horn H M, Starrfield S G and Truran J W 1977 *Astrophys. J.* **212** 791
Conti P S 1976a *IAU Symp. No 70, Be and Shell Stars* ed A Slettebak (Dordrecht: Reidel) p446
—— 1976b *Mem. Soc. R. Sci. Liege* series 6 **9** 193
—— 1978 *Ann. Rev. Astron. Astrophys.* **16** 371
—— 1979 *IAU Symp. No 83, Mass Loss and Evolution of O Type Stars* ed P S Conti and C W H de Loore (Dordrecht: Reidel) p431
Conti P S and Garmany C D 1980 *Astrophys. J.* **238** 190
Conti P S, Massey P, Ebbets D and Niemela V S 1980 *Astrophys. J.* **238** 184
Conti P S and Niemela V S 1976 *Astrophys. J.* **209** L37
Cordova F A, Chester T S, Tuohy I R and Garmire G P 1980a *Astrophys. J.* **235** 163
Cordova F A and Garmire G P 1979 *Nature* **279** 782
Cordova F A and Mason K O 1980 *Nature* **287** 25
Cordova F A, Nugent J J, Klein S R and Garmire G P 1980b *Mon. Not. R. Astron. Soc.* **190** 87
Cordova F A and Riegler G R 1979 *Mon. Not. R. Astron. Soc.* **188** 103
Cowley A P, Crampton D and Hutchings J B 1980 *Astrophys. J.* **241** 269
Coyne G V 1976 *IAU Symp. No 70, Be and Shell Stars* ed A Slettebak (Dordrecht: Reidel) p233
Crampton D, Cowley A P and Hutchings J B 1980 *Astrophys. J.* **235** L131
Crampton D, Hutchings J B and Cowley A P 1979 *Astrophys. J.* **234** 182
Crawford D L, Barnes J V and Perry C L 1975 *Publ. Astron. Soc. Pacific* **87** 115
Cruise A M 1977 *Nature* **267** 685
Dickel H R, Habing H J and Isaacman R 1980 *Astrophys. J.* **238** L39
Dinnerstein H L 1980 *Astrophys. J.* **237** 486

Doazan V 1965 *Annls Astrophys.* **28** 1
—— 1976 *IAU Symp. No 70, Be and Shell Stars* ed A Slettebak (Dordrecht: Reidel) p37
Doazan V, Kuhi L V and Thomas R N 1980 *Astrophys. J.* **235** L17
Dodson H W 1936 *Astrophys. J.* **84** 180
Drake S A and Ulrich R K 1980 *Astrophys. J. Suppl.* **42** 351
Durney B and Roberts P 1971 *Astrophys. J.* **170** 319
Eaton J A 1978 *Astrophys. J.* **220** 582
Eddington A S 1959 *Internal Constitution of the Stars* (New York: Dover)
Edwards D L 1944 *Mon. Not. R. Astron. Soc.* **104** 283
Enmis D, Becklin E, Beckwith S, Elias J, Gatley I, Matthews K, Neugebauer G and Willner S P 1977 *Astrophys. J.* **214** 478
Fabbiano G, Gursky H, Schwartz D A, Schwartz J, Bradt H V and Doxey R E 1978 *Nature* **275** 721
Fabian A C 1980 *Mon. Not. R. Astron. Soc.* **192** 11P
Fabian A C and Hansen C J 1979 *Mon. Not. R. Astron. Soc.* **187** 283
Fabian A C and Pringle J E 1977 *Mon. Not. R. Astron. Soc.* **180** 749
Fabian A C, Pringle J E, Stickland D J and Whelan J A J 1980 *Mon. Not. R. Astron. Soc.* **191** 457
Fabian A C and Rees M J 1980 *Mon. Not. R. Astron. Soc.* **187** 13P
Fahlman G G and Walker G A H 1980 *Astrophys. J.* **240** 169
Fang L Z 1981 *Mon. Not. R. Astron. Soc.* **195** 177
Faraggiana R 1969 *Astron. Astrophys.* **2** 162
Feinstein A and Marraco H G 1979 *Astron. J.* **84** 1713
Ferch R L and Salpeter E E 1975 *Astrophys. J.* **202** 195
Ferland G J 1979a *Astrophys. J.* **231** 781
—— 1979b *Mon. Not. R. Astron. Soc.* **188** 669
Ferland G J, Lambert D L, Netzer H, Hall D N B and Ridgway S T 1979 *Astrophys. J.* **227** 489
Ferland G J and Shields G A 1978a *Astrophys. J.* **224** L15
—— 1978b *Astrophys. J.* **226** 172
Ferland G J and Truran T W 1980 *Astrophys. J.* **240** 608
Ferland G J and Wootten H A 1977 *Astrophys. J.* **214** L27
Ferrari-Toniolo M, Persi P and Viotti R 1978 *Mon. Not. R. Astron. Soc.* **185** 841
Fiebelman W A, Boggess A, Hobbs R W and McCracken C W 1980 *Astrophys. J.* **241** 725
Fiebelman W A, Hobbs R W, McCracken C W and Brown L W 1979 *Astrophys. J.* **231** 111
Fitzgerald M P and Pilavaki A 1974 *Astrophys. J. Suppl.* **28** 147
Flannery B P 1975 *Astrophys. J.* **201** 661
Flora U and Hack M 1975 *Astron. Astrophys. Suppl.* **19** 57
Florkowski D R and Gottesman S T 1977 *Mon. Not. R. Astron. Soc.* **179** 105
Flower D R 1980 *Mon. Not. R. Astron. Soc.* **193** 511
Flower D R and Perinotto M 1980 *Mon. Not. R. Astron. Soc.* **191** 301
Ford H C 1977 *Astrophys. J.* **219** 437
Forrest W J, McCarthy J F and Houck J R 1980 *Astrophys. J.* **240** L37
de Freitas-Pacheco J A 1977 *Mon. Not. R. Astron. Soc.* **181** 421

de Freitas-Pacheco J A 1979 *Mon. Not. R. Astron. Soc.* **186** 617
Frontera F, Fuligni F, Morelli E and Ventura G 1979 *Astrophys. J.* **229** 291
Furenlid I and Young A 1980 *Astrophys. J.* **240** 159
Gallagher J S 1977 *Astron. J.* **82** 209
Gallagher J S, Hege E K, Kopriva D A, Williams R E and Butcher H R 1980 *Astrophys. J.* **237** 55
Gallagher J S and Starrfield S 1976 *Mon. Not. R. Astron. Soc.* **176** 53
—— 1978 *Ann. Rev. Astron. Astrophys.* **16** 171
Geisel S L, Kleinman D E and Low F J 1970 *Astrophys. J.* **161** L101
George D, Kaftan-Kassim M A and Hartsuijker A P 1974 *Astron. Astrophys.* **35** 219
George V and Coyne S J 1974 *Mon. Not. R. Astron. Soc.* **169** 7P
Gerhz R D, Grasdalen G L and Hackwell J A 1980a *Astrophys. J.* **237** 855
Gerhz R D and Hackwell J A 1974 *Astrophys. J.* **194** 619
Gerhz R D, Hackwell J A, Grasdalen G L, Ney E P, Neugebauer G and Sellgren K 1980b *Astrophys. J.* **239** 570
Gerhz R D, Hackwell J A and Jones T W 1974 *Astrophys. J.* **191** 675
Giles A B, King A R, Cooke B A, McHardy I M and Lawrence A 1979 *Nature* **281** 282
Giles A B, King A R, Jameson R F, Sherrington M R, Hough J H, Bailey J and Cunningham E C 1980 *Nature* **286** 689
Gillett F C and Stein W A 1971 *Astrophys. J.* **164** 77
Giuricin G, Mardirossian F, Mezzetti M, Pucillo M, Santin P and Sedmak G 1979 *Astron. Astrophys.* **80** 9
Glasby J S 1970 *The Dwarf Novae* (London: Constable)
Glass I S 1979 *Mon. Not. R. Astron. Soc.* **187** 305
Gopal-Krishna 1978 *Mon. Not. R. Astron. Soc.* **182** 723
Goudis C, McMullen D, Meaburn J, Tebbutt C and Terrett D C 1978 *Mon. Not. R. Astron. Soc.* **182** 13
Grasdalen G L 1979 *Astrophys. J.* **229** 587
Gray D F 1973 *Astrophys. J.* **184** 461
Greenstein J L 1960 *Stars and Stellar Systems* **6** ed J L Greenstein (University of Chicago Press) p676
Greenstein J L, Sargent W L W, Bordson T A and Boksenberg A 1978 *Astrophys. J.* **218** L121
Gregory P C, Kwok S and Seaquist E 1977 *Astrophys. J.* **211** 429
Greig W E 1971 *Astron. Astrophys.* **10** 161
—— 1972 *Astron. Astrophys.* **18** 70
Griffiths R E, Bradt H, Doxey R, Freidman H, Gursky H, Johnston M, Longmore A, Malin D F, Murdin P, Schwartz D A and Schwartz J 1978 *Astrophys. J.* **221** L63
Griffiths R E, Lamb D Q, Ward M J, Wilson A, Charles P A, Thorstensen J, McHardy I M and Lawrence A 1980 *Mon. Not. R. Astron. Soc.* **193** 25P
Grimin V P and Zvereva A M 1968 *IAU Symp. No 34, Planetary Nebulae* ed D E Osterbrock and C R O'Dell (Dordrecht: Reidel) p278
de Groot M 1969 *Mass Loss from Stars* ed M Hack (Dordrecht: Reidel) p26
—— 1973 *IAU Symp. No 49, Wolf-Rayet and High Temperature Stars* ed M K V Bappu and J Sahade (Dordrecht: Reidel) p108

Grouillier H 1935 *Lyons Observatory Bull.* **1** 16
Gull T R and Sofia S 1979 *Astrophys. J.* **230** 782
Gurzadyan G A 1970 *Planetary Nebulae* (Dordrecht: Reidel)
—— 1978 *IAU Symp. No 76, Planetary Nebulae – Observation and Theory* ed Y Terzian (Dordrecht: Reidel) p79
Hack M and Struve O 1971 *Stellar Spectroscopy; Peculiar Stars* (Trieste: Obs. Astron. Trieste)
Hackwell J A, Gerhz R D and Grasdalen G L 1979 *Astrophys. J.* **234** 133
Hackwell J A, Gerhz R D and Smith J R 1974 *Astrophys. J.* **192** 383
Haefner R, Schoembs R and Vogt N 1977 *Astron. Astrophys.* **61** L37
—— 1979 *Astron. Astrophys.* **77** 7
Haisch B M and Cassinelli J P 1976 *Astrophys. J.* **208** 253
Hamann W-R 1980 *Astron. Astrophys.* **84** 342
Hammerschlag-Hensberg G 1978 *Astron. Astrophys.* **64** 399
Hammerschlag-Hensberg G, van den Heuvel E P J, Lamers H J G L M, Burger M, de Loore C, Glencross W, Howarth I, Willis A J, Wilson R, Menzies J, Whitlock P A, van Dessel E L and Sanford P 1980 *Astron. Astrophys.* **85** 119
Hammerschlag-Hensberg G and Wu C-C 1977 *Astron. Astrophys.* **56** 433
Hanbury-Brown R, Davis J, Herbison-Evans D and Allen L R 1970 *Mon. Not. R. Astron. Soc.* **148** 103
Harmanec P and Kriz S 1976 *IAU Symp. No 70, Be and Shell Stars* ed A Slettebak (Dordrecht: Reidel) p385
Harmon R J and Seaton M J 1966 *Mon. Not. R. Astron. Soc.* **132** 15
Harnden F R Jr, Fabricant D, Topka K, Flannery B P, Tucker W N and Gorenstein P 1977 *Astrophys. J.* **214** 418
Harrington J P, Lutz J H, Seaton M J and Stickland D J 1980 *Mon. Not. R. Astron. Soc.* **191** 13
Hartmann L 1978a *Astrophys. J.* **221** 193
—— 1978b *Astrophys. J.* **224** 520
Hartmann L, Jaffe D and Huchra J P 1980 *Astrophys. J.* **239** 905
Hartwick F D A and Hutchings J B 1978 *Astrophys. J.* **226** 203
Harvey P M and Lada C J 1980 *Astrophys. J.* **237** 61
Havlen R J and Moffat A F J 1977 *Astron. Astrophys.* **58** 351
Hayes D P 1978 *Astrophys. J.* **219** 952
Heap S R 1977a *Astrophys. J.* **215** 864
—— 1977b *Astrophys. J.* **217** 90
Hearn A G 1979 *IAU Symp. No 83, Mass Loss and Evolution of O Type Stars* ed P S Conti and C W H de Loore (Dordrecht: Reidel) p169
Heck A, Burki G, Bianchi L, Cassatella A and Clavel J 1980 *Mon. Not. R. Astron. Soc.* **192** 59P
Heise J, Mewe R, Brinkman A C, Gronenschild E H B M, de Boggende A F J, Schrijver J, Parsignault D R and Grindley J E 1978 *Astron. Astrophys.* **63** L1
Heligman G M 1980 *Mon. Not. R. Astron. Soc.* **191** 761
Hendry E M 1976 *IAU Symp. No 70, Be and Shell Stars* ed A Slettebak (Dordrecht: Reidel) p429
Henize K G 1976 *Astrophys. J. Suppl.* **30** 491

Henize K G, Wray J D, Parsons S B and Benedict G F 1975 *Astrophys. J.* **199** L173

Herbig G H 1960 *Astrophys. J. Suppl.* **4** 337

Herbst W, Nesseu J E and Ostriker J P 1974 *Astrophys. J.* **193** 679

van den Heuvel E P J 1976 *IAU Symp. No 73, Structure and Evolution of Close Binary Systems* ed P Eggleton, S Mitton and J Whelan (Dordrecht: Reidel) p35

van den Heuvel E P J, Ostriker J P and Petterson J A 1980 *Astron. Astrophys.* **81** L7

Hildebrand R H, Spillar E J, Middleditch J, Patterson J J and Steining R F 1980 *Astrophys. J.* **238** L145

Hiltner W A and Schild R E 1966 *Astrophys. J.* **143** 770

Hjellming R M 1976 *Physics of Non-Thermal Radio Sources* ed G Setti (Dordrecht: Reidel) p203

Hjellming R M, Wade C M, Vandenberg N R and Newell R T 1979 *Astron. J.* **84** 1619

Hjellming R M, Webster E and Bauck B 1972 *Astrophys. J.* **178** L139

Holm A V and Gallagher J S III 1974 *Astrophys. J.* **192** 425

van Horn H M, Wesemael F and Winget D E 1980 *Astrophys. J.* **235** L143

Horne K and Gomer R 1980 *Astrophys. J.* **237** 845

Huang S-S 1973 *Astrophys. J.* **183** 541

—— 1975 *Sky and Telescope* **49** 359

—— 1976 *Publ. Astron. Soc. Pacific* **88** 448

—— 1977 *Astrophys. J.* **212** 123

—— 1978 *Astrophys. J.* **219** 956

Huber M C E, Nussbaumer H, Smith L J, Willis A J and Wilson R 1979 *Nature* **278** 697

Hummer D G 1976 *IAU Symp. No 70, Be and Shell Stars* ed A Slettebak (Dordrecht: Reidel) p281

Hummer D G and Rybicki G B 1967 *Methods in Computational Physics* **7** 53

—— 1971 *Ann. Rev. Astron. Astrophys.* **9** 237

Humphreys R M 1978 *IAU Symp. No 80, The H-R Diagram* ed A G Davis-Phillip and D S Hayes (Dordrecht: Reidel) p263

Hutchings J B 1969 *Non-periodic Phenomena in Variable Stars* ed L Detre (Dordrecht: Reidel) p191

—— 1970a *Mon. Not. R. Astron. Soc.* **150** 55

—— 1970b *Stellar Rotation* ed A Slettebak (Dordrecht: Reidel) p283

—— 1974 *IAU Symp. No 59, Stellar Instability and Evolution* ed P Ledoux, A Noels and A W Rodgers (Dordrecht: Reidel) p127

—— 1976a *Astrophys. J.* **204** L99

—— 1976b *IAU Symp. No 70, Be and Shell Stars* ed A Slettebak (Dordrecht: Reidel) p13

—— 1977 *Mon. Not. R. Astron. Soc.* **181** 619

—— 1979a *Astrophys. J.* **232** 176

—— 1979b *Astrophys. J.* **233** 913

Hutchings J B, Cowley A P and Crampton D 1979 *Astrophys. J.* **232** 500

Hutchings J B, Cowley A P and Redman R O 1975 *Astrophys. J.* **201** 404

Hutchings J B and McCall M L 1977 *Astrophys. J.* **217** 775

Hutchings J B and von Rudloff I R 1980 *Astrophys. J.* **238** 909
Ilovaisky S A, Chevalier C, White N E, Mason K O, Sanford P W, Delvaille J P and Schnopper H W 1980 *Mon. Not. R. Astron. Soc.* **191** 81
Jacoby G H 1980 *Astrophys. J. Suppl.* **42** 1
Jameson R F, Akinci R, Adams D J, Giles A B and McCall A 1978 *Nature* **271** 334
Jameson R F, King A R and Sherrington M R 1980 *Mon. Not. R. Astron. Soc.* **191** 559
Jaschek C, Ferrer L and Jaschek M 1971 *Obs. Astron. Univ. Natl. La Plata Ser. Astron.* **37**
Jaschek M, Hubert-Delplace A-M, Hubert H and Jaschek C 1980 *Astron. Astrophys. Suppl.* **42** 103
Johnson H M 1976 *Astrophys. J.* **206** 469
—— 1978 *Astrophys. J. Suppl.* **36** 217
Jones T J 1979 *Astrophys. J.* **228** 787
Kaler J B 1976 *Astrophys. J.* **210** 843
—— 1979 *Astrophys. J.* **228** 163
—— 1980 *Astrophys. J.* **239** 592
Kaler J B, Aller L H, Czyzak S J and Epps H W 1976 *Astrophys. J. Suppl.* **31** 163
Kalkofen W 1970 *National Bureau Standards Special Publ. no* 332 p120
Katz J I 1975 *Astrophys. J.* **200** 298
Kemp J C, Barbour M S, Arbabi M, Leibowitz E M and Mazeh T 1980 *Astrophys. J.* **238** L133
Kemp J C, Swedlund J B and Wolstencroft R 1974 *Astrophys. J.* **193** L15
Kemp J C, Sykes M V and Rudy R J 1977 *Astrophys. J.* **211** L71
Khromov G S 1968 *IAU Symp. No 34, Planetary Nebulae* ed D E Osterbrock and C R O'Dell (Dordrecht: Reidel) p330
King A R and Lasota J P 1979 *Mon. Not. R. Astron. Soc.* **188** 653
King A R, Ricketts M J and Warwick R S 1979 *Mon. Not. R. Astron. Soc.* **187** 77P
Kiplinger A L 1979 *Astron. J.* **84** 655
—— 1980 *Astrophys. J.* **236** 839
Kitchin C R 1970a *Astrophys. Space Sci.* **8** 3
—— 1970b *Mon. Not. R. Astron. Soc.* **150** 455
—— 1971 *PhD Thesis* University of Leicester
—— 1973a *Mon. Not. R. Astron. Soc.* **161** 381
—— 1973b *Mon. Not. R. Astron. Soc.* **161** 389
—— 1976 *Astrophys. Space Sci.* **45** 119
—— 1982 *Mon. Not. R. Astron. Soc.* **198**
Kitchin C R and Meadows A J 1970 *Astrophys. Space Sci.* **8** 463
Kleine T and Kohoutek L 1979 *Astron. Astrophys.* **76** 133
Kogure T, Hirata R and Asada Y 1978 *Publ. Astron. Soc. Japan* **30** 385
Kohoutek L and Lausten S 1977 *Astron. Astrophys.* **61** 761
Kopal Z 1978 *Dynamics of Close Binary Systems* Astrophysics and Space Science Library Vol 68) (Dordrecht: Reidel)
Koppen J and Tarafdar S P 1978 *Astron. Astrophys.* **69** 363
Koppen J and Wehrse R 1980a *Astron. Astrophys.* **85** L15
—— 1980b *Report of 2nd European IUE Conf.* ed B Battrick and J Mort (European Space Agency Sp57) 191

Kraft R P and Luyten W J 1965 *Astrophys. J.* **142** 1041
Kriz S and Harmanec P 1975 *Bull. Astron. Inst. Czechoslovakia* **26** 65
Kuan P and Kuhi L V 1975 *Astrophys. J.* **199** 148
Kuhi L V 1973 *IAU Symp. No 49, Wolf-Rayet and High Temperature Stars* ed M K V Bappu and J Sahade (Dordrecht: Reidel) p205
Kunasz P B 1980 *Astrophys. J.* **237** 819
Kundt W 1979 *Nature* **282** 52
Kwok S 1980 *Astrophys. J.* **236** 592
Kwok S and Purton C R 1979 *Astrophys. J.* **229** 187
Kwok S, Purton C R and Fitzgerald P M 1978 *Astrophys. J.* **219** L125
Lacy C H 1977 *Astrophys. J.* **212** 132
Lambert D L and Tomkin J 1979 *Astrophys. J.* **228** L37
Lamers H J G L M and Morton D C 1976 *Astrophys. J. Suppl.* **32** 715
Lamers H J G L M, Paerels F B S and de Loore C 1980 *Astron. Astrophys.* **87** 68
Lamers H J G L M and Rogerson J B 1978 *Astron. Astrophys.* **66** 417
Lang K R 1978 *Astrophys. Formulae* (Berlin: Springer)
Leep E M 1978 *Astrophys. J.* **225** 165
—— 1979 *IAU Symp. No 83, Mass Loss and Evolution of O Type Stars* ed P S Conti and C W H de Loore (Dordrecht: Reidel) p471
Leep E M and Conti P S 1979 *Astrophys. J.* **228** 224
Liebert J, Angel J R P, Hege E K, Martin P G and Blair W P 1979 *Nature* **279** 384
Liebert J and Stockman H S 1979 *Astrophys. J.* **229** 652
Liebert J, Stockman H S, Angel J R P, Woolf N J, Hege K and Margon B 1978 *Astrophys. J.* **225** 201
Liller W 1975 *IAU Circular No* 2780
—— 1978 *IAU Symp. No 76, Planetary Nebulae – Observation and Theory* ed Y Terzian (Dordrecht: Reidel) p35
Limber D N 1974 *Astrophys. J.* **192** 429
—— 1976 *IAU Symp. No 70, Be and Shell Stars* ed A Slettebak (Dordrecht: Reidel) p371
Lin D N C and Pringle J E 1976 *IAU Symp. No 73, Structure and Evolution of Close Binary Systems* ed P Eggleton, S Mitton and J Whelan (Dordrecht: Reidel) p237
Livio M, Salzman J and Shaviv G 1979 *Mon. Not. R. Astron. Soc.* **188** 1
Lloyd C, Noble R and Penston M V 1977 *Mon. Not. R. Astron. Soc.* **179** 675
Lloyd-Evans T 1980 *Mon. Not. R. Astron. Soc.* **192** 47
Lo K Y and Bechis K P 1976 *Astrophys. J.* **205** L21
Long K S and Kestenbaum H L 1978 *Astrophys. J.* **226** 271
Long K S and White R L 1980 *Astrophys. J.* **239** L65
Lortet M C, Niemela V S and Tarsia R 1980 *Astron. Astrophys.* **90** 210
Losh H M 1931 *Publ. Michigan Observatory* **4** 1
Lucy L B and Solomon P 1970 *Astrophys. J.* **159** 879
Lucy L B and White R L 1980 *Astrophys. J.* **241** 300
Lutz J H 1974 *Publ. Astron. Soc. Pacific* **86** 888
—— 1977 *Astron. Astrophys.* **60** 93
—— 1978 *IAU Symp. No 76, Planetary Nebulae – Observation and Theory* ed Y Terzian (Dordrecht: Reidel) p185
Lutz J H and Carnochan D J 1979 *Mon. Not. R. Astron. Soc.* **189** 701

Lutz J H and Seaton M J 1979 *Mon. Not. R. Astron. Soc.* **187** 1P
Luud L S 1967 *Sov. Astron.* **11** 211
McAlary C W and McLaren R A 1980 *Astrophys. J.* **240** 853
McCarthy J F, Forrest W J and Houck J R 1978 *Astrophys. J.* **224** 109
MacDonald J 1980 *Mon. Not. R. Astron. Soc.* **191** 933
MacGregor A D, Sanchez-Magro C. Selby M J and Whitelock P A 1976 *Astron. Astrophys.* **50**, 389
McLaughlin D B 1937 *Astrophys. J.* **85** 181
—— 1960 *Stars and Stellar Systems* **6** ed J L Greenstein (Chicago Press) p585
—— 1961 *J. R. Astron. Soc. Canada* **55** 73
—— 1963 *Astrophys. J.* **137** 1085
McLean I S 1979 *Mon. Not. R. Astron. Soc.* **186** 265
—— 1980 *Astrophys. J.* **236** L149
McLean I S and Clarke D 1979 *Mon. Not. R. Astron. Soc.* **186** 245
McLean I S, Coyne G V, Frecker J E and Serkowski K 1979a *Astrophys. J.* **228** 802
—— 1979b *Astrophys. J.* **231** L141
Maciel W J and Pottasch S R 1980 *Astron. Astrophys.* **88** 1
Maeder A, Lequeux L and Azzopardi M 1980 *Astron. Astrophys.* **90** L17
Maeder A and Rufener F 1974 *IAU Symp. No 59, Stellar Instability and Evolution* ed P Ledoux, A Noels and A W Rodgers (Dordrecht: Reidel) p81
Mardirossian F, Mezzetti M, Pucillo M, Santin P. Sedmak G and Giuricin G 1980 *Astron. Astrophys.* **85** 29
Margon B and Abell G 1979 *Nature* **279** 701
Margon B, Bowyer S and Penegor G 1976 *Mon. Not. R. Astron. Soc.* **176** 217
Margon B, Ford H C, Grandi S A and Stone R P S 1979a *Astrophys. J.* **233** L63
Margon B, Ford H C, Katz J I, Kwitter K B, Ulrich R K, Stone R P S and Klemola A 1979b *Astrophys. J.* **230** L41
Margon B, Grandi S A and Downes R A 1980 *Astrophys. J.* **241** 306
Marlborough J M 1969 *Astrophys. J.* **156** 135
—— 1970 *Astrophys. J.* **159** 575
—— 1976 *IAU Symp. No 70, Be and Shell Stars* ed A Slettebak (Dordrecht: Reidel) p335
—— 1977 *Astrophys. J.* **216** 446
Marlborough J M and Cowley A P 1974 *Astrophys. J.* **187** 99
Marlborough J M and Roy J R 1970 *Astrophys. J.* **160** 221
—— 1971 *Astrophys. J.* **169** 327
Marlborough J M and Snow T P 1976 *IAU Symp. No 70, Be and Shell Stars* ed A Slettebak (Dordrecht: Reidel) p179
Martin P G and Rees M J 1979 *Mon. Not. R. Astron. Soc.* **189** 19P
Mason K O, Kahn S M and Bowyer C S 1979 *Nature* **280** 568
Mason K O, Lampton M, Charles P and Bowyer C S 1978a *Astrophys. J.* **226** L129
Mason K O, Murdin P G, Parkes G E and Visvanathan N 1978b *Mon. Not. R. Astron. Soc.* **184** 45P
Mason K O, White N E and Sanford P W 1976 *Nature* **260** 690
Massey P 1980 *Astrophys. J.* **236** 526
Massey P and Conti P S 1977 *Astrophys. J.* **218** 431
Mathews W G 1969 *Astrophys. J.* **157** 583

Mayall M W 1969 *J. R. Astron. Soc. Canada* **63** 221
Mayo S K, Wickramasinghe D T and Whelan J A J 1980 *Mon. Not. R. Astron. Soc.* **193** 793
Melnick G and Harwit M 1975 *Mon. Not. R. Astron. Soc.* **171** 441
Mendez R H 1978 *Mon. Not. R. Astron. Soc.* **185** 647
Mendez R H and Niemela V S 1977 *Mon. Not. R. Astron. Soc.* **178** 409
Mendez R H, Niemela V S and Lee P 1978 *Mon. Not. R. Astron. Soc.* **184** 351
Mendoza E E 1958 *Astrophys. J.* **128** 207
Merrill P W and Burwell C G 1933 *Astrophys. J.* **78** 87
—— 1943 *Astrophys. J.* **98** 153
—— 1949 *Astrophys. J.* **110** 387
—— 1950 *Astrophys. J.* **112** 72
Mihalas D 1978 *Stellar Atmospheres* (San Francisco: W H Freeman)
—— 1979 *Mon. Not. R. Astron. Soc.* **189** 671
Milgrom M 1979a *Astron. Astrophys.* **76** L3
—— 1979b *Astron. Astrophys.* **78** L9
—— 1980 *Astron. Astrophys.* **87** L15
Miller J S 1974 *Ann. Rev. Astron. Astrophys.* **12** 331
Moffat A F J 1977 *Astron. Astrophys.* **57** 151
Moffat A F J and Isserstedt J 1980 *Astron. Astrophys.* **85** 201
Moffat A F J and Seggewis W 1979 *IAU Symp. No 83, Mass Loss and Evolution of O Type Stars* ed P S Conti and C W H de Loore (Dordrecht: Reidel) p447
Moffett T J and Barnes T G III 1974 *Astrophys. J.* **194** 141
Mollenhoff C and Schaifers K 1978 *Astron. Astrophys.* **64** 253
Morrison N D and Conti P S 1978 *Astrophys. J.* **224** 558
Morton D C 1967 *Astrophys. J.* **147** 1017
Morton D C and Wright A E 1978 *Mon. Not. R. Astron. Soc.* **182** 47P
Moseley H 1980 *Astrophys. J.* **238** 892
Moss D 1977 *Mon Not. R. Astron. Soc.* **178** 61
Murdin P 1975 *IAU Circular No* 2784
Murdin P, Clark D H and Martin P G 1980 *Mon. Not. R. Astron. Soc.* **193** 135
Murdin P, Griffiths R E, Pounds K A, Watson M G and Longmore A 1977 *Mon. Not. R. Astron. Soc.* **178** 27P
Murdin P, Morton D C and Thomas R N 1979 *Mon. Not. R. Astron. Soc.* **186** 43P
Mutson S L, Lyon J and Marionni P A 1975 *Astrophys. J.* **201** L85
Nandy K, Morgan D H, Willis A J and Gondhalekar P 1980 *Mon. Not. R. Astron. Soc.* **193** 43P
Nather R E and Robinson E L 1974 *Astrophys. J.* **190** 637
Nevo I and Sadeh D 1976 *Mon. Not. R. Astron. Soc.* **177** 167
—— 1978 *Mon. Not. R. Astron. Soc.* **182** 595
New Scientist 1980 **88** 228
Ney E and Hatfield B F 1978 *Astrophys. J.* **219** L111
Niemela V S 1979 *IAU Symp. No 83, Mass Loss and Evolution of O Type Stars* ed P S Conti and C W H de Loore (Dordrecht: Reidel) p475
Niemela V S and Sahade J 1980 *Astrophys. J.* **238** 244
Noerdlinger P D 1979 *IAU Symp. No 83, Mass Loss and Evolution of O Type Stars* ed P S Conti and C W H de Loore (Dordrecht: Reidel) p253

Nugis T, Kolka I and Luud L 1979 *IAU Symp. No 83, Mass Loss and Evolution of O Type Stars* ed P S Conti and C W H de Loore (Dordrecht: Reidel) p39
Nussbaumer H, Schmutz W, Smith L J, Willis A and Wilson R 1979 *Symp. Report on 1st Year of IUE* p259
Oergerle W R and van Blerkom D 1976 *Astrophys. J.* **208** 453
Ortolani S, Rafanelli P, Rosino L and Vittone A 1980 *Astron. Astrophys.* **87** 31
Osterbrock D E 1974 *Astrophysics of Gaseous Nebulae* (San Francisco: W H Freeman)
Paczynski B 1971 *Ann. Rev. Astron. Astrophys.* **9** 183
—— 1973 *IAU Symp. No 49, Wolf–Rayet and High Temperature Stars* ed M K V Bappu and J Sahade (Dordrecht: Reidel) p143
Paczynski B and Rudak B 1980 *Astron. Astrophys.* **82** 349
Papaloizou J and Pringle J E 1978a *Astron. Astrophys.* **70** L65
—— 1978b *Mon. Not. R. Astron. Soc.* **182** 423
—— 1979 *Mon. Not. R. Astron. Soc.* **189** 293
—— 1980 *Mon. Not. R. Astron. Soc.* **190** 13P
van Paradijs J A, Hammerschlag-Hensberg G and Zuiderwijk E J 1978 *Astron. Astrophys. Suppl.* **31** 189
Parker E 1958 *Astrophys. J.* **128** 664
Parkes, G E, Murdin P G and Mason K O 1980 *Mon. Not. R. Astron. Soc.* **190** 537
Patterson J 1979a *Astrophys. J.* **233** L13
—— 1979b *Astrophys. J.* **234** 978
—— 1980 *Astrophys. J.* **241** 235
Patterson J, Branch D, Chincarini G and Robinson E L 1980 *Astrophys. J.* **240** L133
Patterson J, Nather R E, Robinson E L and Handler F 1979 *Astrophys. J.* **232** 819
Patterson J, Robinson E L and Kiplinger A L 1978 *Astrophys. J.* **226** L137
Patterson J, Robinson E L and Nather R E 1977 *Astrophys. J.* **214** 144
Payne-Gaposhkin C 1957 *The Galactic Novae* (Amsterdam: North Holland)
—— 1958 *Handbuch Phys.* **51** 752
—— 1977 *Novae and Related Stars* ed M Freidjung (Dordrecht: Reidel) p3
Peimbert M 1978 *IAU Symp. No 76, Planetary Nebulae – Observation and Theory* ed Y Terzian (Dordrecht: Reidel) p215
Perek L and Kohoutek L 1967 *Catalogue of Galactic Planetary Nebulae* (Prague: Academia)
Perinotto M 1975 *Astron. Astrophys.* **39** 383
Persi P, Viotti R and Ferrari-Toniolo M 1977 *Mon. Not. R. Astron. Soc.* **181** 685
Petterson J A 1980 *Astrophys. J.* **241** 247
Phillips J P and Reay N K 1977 *Astron. Astrophys.* **59** 91
Phillips J P and Selby M J 1977 *Astrophys. Space Sci.* **49** 339
Phillips J P, Selby M L, Wade R and Sanchez-Magro C 1980 *Mon. Not. R. Astron. Soc.* **190** 377
Phillips J P, Wade R, Selby M J and Sanchez-Magro C 1979 *Mon. Not. R. Astron. Soc.* **187** 45P
Piirola V 1979 *Astron. Astrophys. Suppl.* **38** 193
Plavec M 1976 *IAU Symp. No 70, Be and Shell Stars* ed A Slettebak (Dordrecht: Reidel) p439

Poeckert R 1975 *Astrophys. J.* **196** 777
Poeckert R and Marlborough J M 1976 *Astrophys. J.* **206** 182
—— 1978 *Astrophys. J. Suppl.* **38** 229
Pottasch S R 1980 *Astron. Astrophys.* **89** 336
Pottasch S R, Wesselius P R, Wu C-C and van Duinen R J 1978 *Astron. Astrophys.* **62** 95
Prialnik D, Shara M M and Shaviv G 1978 *Astron. Astrophys.* **62** 339
—— 1979 *Astron. Astrophys.* **72** 192
Priedhorsky W C, Krzeminski W and Tapia S 1978a *Astrophys. J.* **225** 542
Priedhorsky W, Matthews K, Neugebauer G, Werner M and Krzeminski W 1978b *Astrophys. J.* **226** 397
Pringle J E 1977 *Mon. Not. R. Astron. Soc.* **178** 195
Purgathofer A and Weinberger R 1980 *Astron. Astrophys.* **87** L5
Purton C 1976 *IAU Symp. No 70, Be and Shell Stars* ed A Slettebak (Dordrecht: Reidel) p157
Rank D M 1978 *IAU Symp. No 76, Planetary Nebulae – Observation and Theory* ed Y Terzian (Dordrecht: Reidel) p103
Rappaport S, Joss P C, Bradt H, Clark G W and Jernigan J G 1976 *Astrophys. J.* **208** L119
Rappaport S, Li S, Clark G W and Jernigan J G 1979 *Astrophys. J.* **228** 893
Raymond J C, Black J H, Davis R J, Dupree A K, Gursky H, Hartmann L and Matilsky T A 1979 *Astrophys. J.* **230** L95
Ricketts M J, King A R and Raine D J 1979 *Mon. Not. R. Astron. Soc.* **186** 233
Ringuelet A E 1980 *Mon. Not. R. Astron. Soc.* **192** 399
Ritter H 1976 *Mon. Not. R. Astron. Soc.* **175** 279
—— 1980a *Astron. Astrophys.* **85** 362
—— 1980b *Astron. Astrophys.* **86** 204
Roberts M S 1962 *Astron. J.* **67** 79
Robertson J A 1980 *Mon. Not. R. Astron. Soc.* **192** 263
Robinson E L 1974 *Astrophys. J.* **193** 191
—— 1975 *Astron. J.* **80** 515
—— 1976a *Ann. Rev . Astron. Astrophys.* **14** 119
—— 1976b *Astrophys. J.* **203** 485
Robinson E L and Nather R E 1979 *Astrophys. J. Suppl.* **39** 461
Robinson E L, Nather R E and Kiplinger A 1974 *Publ. Astron. Soc. Pacific* **86** 401
Robinson E L, Nather R E and Patterson J 1978 *Astrophys. J.* **219** 168
Rogerson J B and Lamers H J G L M 1975 *Nature* **256** 19
Rose W K and Scott E H 1976 *Astrophys. J.* **204** 516
Rosseland S 1929 *Mon. Not. R. Astron. Soc.* **89** 49
Rottenberg J A 1952 *Mon. Not. R. Astron. Soc.* **112** 125
Roxburgh I W and Strittmatter P A 1965 *Z. Astrophys.* **63** 15
Russell R W, Soifer B T and Willner S P 1977 *Astrophys. J.* **217** L149
—— 1978 *Astrophys. J.* **220** 568
Rybicki G B 1970 *National Bureau Standards Special Publ. No* 332 p87
Ryle M, Caswell J L, Hine G and Shakeshaft J R 1978 *Nature* **276** 571
Sahade J 1980 *Astron. Astrophys.* **87** L7

Sahade J and Wood F B 1978 *Interacting Binary Stars* (Int. Series in Natural Philosophy, Vol 95) (Oxford: Pergamon)
Sanyal A, Weller W and Jeffers S 1974 *Astrophys. J.* **187** L31
Sanyal A and Wilison L A 1980 *Astrophys. J.* **237** 529
Saraph H E and Seaton M J 1980 *Mon. Not. R. Astron. Soc.* **193** 617
Sarazin C L, Begelman M C and Hatchett S P 1980 *Astrophys. J.* **238** L129
Savage B D, Wesselius P, Swings J P and The P S 1978 *Astrophys. J.* **224** 149
Scalo J M and Shields G A 1979 *Astrophys. J.* **228** 521
Schild R E 1966 *Astrophys. J.* **146** 142
—— 1976 *IAU Symp. No 70, Be and Shell Stars* ed A Slettebak (Dordrecht: Reidel) p107
Schild R E, Chaffee F, Frogel J A and Persson S 1974 *Astrophys. J.* **190** 73
Schild R E and Romanshin W 1976 *IAU Symp. No 70, Be and Shell Stars* ed A Slettebak (Dordrecht: Reidel) p31
Schneider D P and Young P 1980 *Astrophys. J.* **240** 871
Schnopper H W, Delvaille J P, Epstein A, Helmken H, Murray S S, Clark G, Jernigan G and Doxey R 1976 *Astrophys. J.* **210** L75
Schuster A 1905 *Astrophys. J.* **21** 1
Schwartz P R and Spencer J H 1977 *Mon. Not. R. Astron. Soc.* **180** 297
Scott P F 1975 *Mon. Not. R. Astron. Soc.* **170** 487
Seaquist E R 1976 *Astrophys. J.* **203** L35
Seaquist E R, Duric N, Israel F P, Spoelstra T A T, Ulich B L and Gregory P C 1978 *Astron. J.* **85** 283
Seaquist E R, Gilmore W, Nelson G J, Payten W J and Slee O B 1980 *Astrophys. J.* **241** L77
Seaquist E R, Gregory P C and Crane P C 1979 *IAU Circular No* 3256
Seaquist E R and Palimaka J 1977 *Astrophys. J.* **217** 781
Secchi A 1866 *Astron. Nachrichten No* 1612 p63
Seggewis W 1974 *Publ. Astron. Soc. Pacific* **86** 670
Seward F D, Forman W R, Giacconi R, Griffiths R E, Harnden F R Jr, Jones C and Pye J P 1979 *Astrophys. J.* **234** L55
Seward F, Grindlay J, Seaquist E and Gilmore W 1980 *Nature* **287** 806
Shaham J 1980 *Astrophys. Lett.* **20** 115
Shajn G and Struve O 1929 *Mon. Not. R. Astron. Soc.* **89** 222
Shara M M 1980 *Astrophys. J.* **239** 581
Shara M M, Prialnik D and Shaviv G 1980 *Astrophys. J.* **239** 586
Sharov A S and Lyuty V M 1976 *IAU Symp. No 70, Be and Shell Stars* ed A Slettebak (Dordrecht: Reidel) p105
Sherrington M R, Lawson P A, King A R and Jameson R F 1980 *Mon. Not. R. Astron. Soc.* **191** 185
Shields G A 1975 *Astrophys. J.* **195** 475
Shields G A and Ferland G J 1978 *Astrophys. J.* **225** 950
Sitko M and Savage B D 1980 *Astrophys. J.* **237** 82
Slettebak A 1969 *Non-periodic Phenomena in Variable Stars* ed L Detre Dordrecht: Reidel) p179
—— 1976 *IAU Symp. No 70, Be and Shell Stars* ed A Slettebak (Dordrecht: Reidel) p123

Slovak M H and Africano J 1978 *Mon. Not. R. Astron. Soc.* **185** 591
Smith L F 1968a *Mon. Not. R. Astron. Soc.* **138** 109
—— 1968b *Mon. Not. R. Astron. Soc.* **140** 409
—— 1973 *IAU Symp. No 49, Wolf–Rayet and High Temperature Stars* ed M K V Bappu and J Sahade (Dordrecht: Reidel) p15
Smith L F and Aller LH 1969 *Astrophys. J.* **157** 1245
—— 1971 *Astrophys. J.* **164** 275
Smith S E 1980 *Astrophys. J.* **237** 831
Sneden C and Lambert D L 1975 *Mon. Not. R. Astron. Soc.* **170** 533
Snow T P and Morton D C 1976 *Astrophys. J. Suppl.* **32** 429
Snow T P, Peters G J and Mathieu R D 1979 *Astrophys. J. Suppl.* **39** 359
Snow T P, Wegner G A and Kunasz P B 1980 *Astrophys. J.* **238** 643
Sobolev V V 1953 *Sov. Astron.* **1** 678
—— 1960 *Moving Envelopes of Stars* (Harvard University Press)
Soderhjelm S 1980 *Astron. Astrophys.* **89** 100
Sparks W M and Kutter G S 1980 *Space Sci. Rev.* **27** 643
Starrfield S 1980 *Space Sci. Rev.* **27** 635
Starrfield S, Sparks W M and Truran J W 1974 *Astrophys. J. Suppl.* **28** 247
—— 1976 *IAU Symp. No 73, Structure and Evolution of Close Binary Systems* ed P Eggleton, S Mitton and J Whelan (Dordrecht: Reidel) p155
Starrfield S, Truran J W and Sparks W M 1979 *Astrophys. J.* **226** 186
Stauffer J, Spinrad H and Thorstensen J 1979 *Publ. Astron. Soc. Pacific* **91** 59
Steavenson W H 1923 *Mon. Not. R. Astron. Soc.* **83** 397
—— 1924 *Mon. Not. R. Astron. Soc.* **84** 538
Stephenson C B 1967 *Publ. Astron. Soc. Pacific* **79** 584
Stephenson C B and Sanduleak N 1977 *Astrophys. J. Suppl.* **33** 459
Stickland D, Penn C J, Seaton M J, Snidjers M A J, Storey P J and Kitchin C R 1979 *Symp. Report on 1st Year of IUE* p63
Stiening R F, Hildebrand R H and Spillar E J 1979 *Publ. Astron. Soc. Pacific* **91** 384
Stockman H S and Sargent T A 1979 *Astrophys. J.* **227** 197
Stoeckley T R 1968 *Mon. Not. R. Astron. Soc.* **140** 141
Strittmatter P A, Woolf N J, Thompson R I, Wilkerson S, Angel J R P, Stockman H S, Gilbert G, Grandi S A, Larson H and Fink U 1977 *Astrophys. J.* **216** 23
Strom S E, Strom K M, Yost J, Carrasco L and Grasdalen G 1972 *Astrophys. J.* **173** 353
Stromgren B 1939 *Astrophys. J.* **89** 526
Struve O 1931 *Astrophys. J.* **73** 94
—— 1944 *Astrophys. J.* **99** 222
—— 1950 *Stellar Evolution* (Princeton University Press)
—— 1951 *Astrophysics – A Topical Symposium* (New York: McGraw-Hill) p85
Surdej J 1979 *Astron. Astrophys.* **73** 1
Swank J H, Becker R H, Boldt E A, Holt S S, Pravdo S H and Serlemitsos P J 1977 *Astrophys. J.* **212** L73
Swank J H, Boldt E A, Holt S S, Rothschild R E and Serlemitsos P J 1978 *Astrophys. J.* **226** L133
Swedlund J B, Kemp J C and Wolstencroft R D 1974 *Astrophys. J.* **193** L11

Swings J P 1973 *Astron. Astrophys.* **26** 443
Swings J P and Allen D A 1971 *Astrophys. J.* **167** 247
Swings J P, Barbier R, Klutz M, Surdej A and Surdej J 1980 *Astron. Astrophys.* **90** 116
Szkody P 1976 *Astrophys. J.* **207** 190
—— 1977 *Astrophys. J.* **217** 140
Szkody P and Margon B 1980 *Astrophys. J.* **236** 862
Tapia S 1977 *Astrophys. J.* **212** L125
Tarasco J D, Wood H D and Roberts M S 1970 *Astrophys. J.* **161** L129
Tayler R J 1980 *Mon. Not. R. Astron. Soc.* **191** 135
Taylor K 1977 *Mon. Not. R. Astron. Soc.* **181** 475
Taylor K and Scarrott S M 1980 *Mon. Not. R. Astron. Soc.* **193** 321
Terlevich R J and Pringle J E 1979 *Nature* **278** 719
Terzian Y 1968 *IAU Symp. No 34, Planetary Nebulae* ed D E Osterbrock and C R O'Dell (Dordrecht: Reidel) p87
—— 1978 *IAU Symp. No 76, Planetary Nebulae – Observation and Theory* ed Y Terzian (Dordrecht: Reidel) p111
Terzian Y, Higgs L A, MacLeod J M and Doherty L H 1974 *Astron. J.* **79** 1018
Thackeray A D 1977a *Mem. R. Astron. Soc.* **83** 1
—— 1977b *Mon. Not. R. Astron. Soc.* **180** 95
—— 1978 *Mon. Not. R. Astron. Soc.* **182** 11P
Thomas R N 1968 *National Bureau of Standards Special Publ. No* 307 p1
Tokunaga A T and Young E T 1980 *Astrophys. J.* **237** L93
Tuchman Y, Sack N and Barkat Z 1978 *Astrophys. J.* **225** L137
—— 1979 *Astrophys. J.* **234** 217
Underhill A B 1966 *The Early Type Stars* (Astrophysics and Space Science Library Vol 6) (Dordrecht: Reidel)
—— 1970 *National Bureau of Standards Special Publ. No* 332 p3
—— 1973 *IAU Symp. No 49, Wolf-Rayet and High Temperature Stars* ed M K V Bappu and J Sahade (Dordrecht: Reidel) p237
—— 1979 *Astrophys. J.* **234** 528
—— 1980 *Astrophys. J.* **239** 220
Vanbeveran D and Doon C 1980 *Astron. Astrophys.* **87** 77
Vidal N V and Wickramasinghe D T 1974 *Astron. Astrophys.* **36** 309
Visvanathan N and Wickramasinghe D T 1979 *Nature* **281** 47
Vogt N 1980 *Astron. Astrophys.* **88** 66
Vogt N and Breysacher J 1980 *Astrophys. J.* **235** 945
Vogt N, Krzeminski W and Sterken C 1980 *Astron. Astrophys.* **85** 106
Vrba F J 1975 *Astrophys. J.* **195** 101
Vrba F J, Schmidt G D and Hintzen P M 1979 *Astrophys. J.* **227** 185
Wackerling L R 1970 *Mem. R. Astron. Soc.* **73** 153
Walborn N R 1971 *Astrophys. J. Suppl.* **23** 257
—— 1974 *Astrophys. J.* **189** 269
—— 1976 *Astrophys. J.* **205** 419
Walker A R 1974 *Mon. Not. R. Astron. Soc.* **169** 47P
—— 1977 *Mon. Not. R. Astron. Soc.* **179** 587

Walker M F 1953 *Astrophys. J.* **118** 481
Warner B 1974 *Mon. Not. R. Astron. Soc.* **168** 235
—— 1975 *IAU Symp. No 73, Structure and Evolution of Close Binary Systems* ed P Eggleton, S Mitton and J Whelan (Dordrecht: Reidel) p85
—— 1980 *Mon. Not. R. Astron. Soc.* **190** 69P
Warner B and Brickhill A J 1978 *Mon. Not. R. Astron. Soc.* **182** 777
Warner B and Robinson E L 1972 *Nature Phys. Sci.* **239** 2
Warner B and Thackeray A D 1975 *Mon. Not. R. Astron. Soc.* **172** 433
Warren W N 1976 *Mon. Not. R. Astron. Soc.* **174** 111
Watson M G, Mayo S K and King A R 1980 *Mon. Not. R. Astron. Soc.* **192** 689
Watson M G, Ricketts M J and Griffiths R E 1978a *Astrophys. J.* **221** L69
Watson M G, Sherrington M R and Jameson R F 1978b *Mon. Not. R. Astron. Soc.* **184** 79P
Watts D J, Greenhill J G and Thomas R N 1980 *Mon. Not. R. Astron. Soc.* **191** 25P
Webster B L 1976 *Mon. Not. R. Astron. Soc.* **174** 513
Webster B L and Glass I S 1974 *Mon. Not. R. Astron. Soc.* **166** 491
Wehmeyer R and Kohoutek L 1979 *Astron. Astrophys.* **78** 39
Weidemann V 1977 *Astron. Astrophys.* **61** L27
Weinberger R 1977 *Astron. Astrophys. Suppl.* **30** 335
Wellman P 1955 *Vistas Astron.* **1** 303
Westerlund B E and Smith L F 1964 *Mon. Not. R. Astron. Soc.* **128** 311
Whelan J A J, Ward M J, Allen D A, Panziger I J, Fosbury R A E, Murdin P G, Penston M V, Peterson B A, Wampler E J and Webster B L 1977 *Mon. Not. R. Astron. Soc.* **180** 657
White N E, Charles P A and Thorstensen J R 1980a *Mon. Not. R. Astron. Soc.* **193** 731
White N E, Holt S S, Becker R H, Boldt E A and Serlemitsos P J 1980b *Astrophys. J.* **239** L69
Whitmire D P and Mateze J J 1980 *Mon. Not. R. Astron. Soc.* **193** 707
—— 1981 *Mon. Not. R. Astron. Soc.* **194** 293
Whitney C A 1978 *Astrophys. J.* **220** 245
Whittet D C B and van Breda I G 1980 *Mon. Not. R. Astron. Soc.* **192** 467
Whyte C A and Eggleton P P 1980 *Mon. Not. R. Astron. Soc.* **190** 801
Williams P M, Adams D J, Arakaki S, Beattie D H, Born J, Lee T J, Robertson D J and Stewart J M 1980 *Mon. Not. R. Astron. Soc.* **192** 25P
Williams P M and Antonopoulou E 1979 *Mon. Not. R. Astron. Soc.* **187** 183
Williams P M, Beattie D H, Lee T J, Stewart J M and Antonopoulou E 1978 *Mon. Not. R. Astron. Soc.* **185** 467
Willis A J 1980 *Report of 2nd European IUE Conf.* ed B Battrick and J Mort (European Space Agency Sp157) il
Willis A J, George D and Kaftan-Kassim M A 1974 *Astron. Astrophys.* **36** 455
Willis A J and Stickland D J 1980 *Mon. Not. R. Astron. Soc.* **190** 27P
Willis A J and Wilson R 1976 *Astron. Astrophys.* **47** 429
—— 1977 *Astron. Astrophys.* **59** 133
—— 1978 *Mon. Not. R. Astron. Soc.* **182** 559
—— 1979 *IAU Symp. No 83, Mass Loss and Evolution of O Type Stars* ed P S

Conti and C W H de Loore (Dordrecht: Reidel) p461
Willner S P, Jones B, Puetter R C, Russell R W and Soiffer B T 1979 *Astrophys. J.* **234** 496
Wolf B and Sterken G 1979 *IAU Symp. No 83, Mass Loss and Evolution of O Type Stars* ed P S Conti and C W H de Loore (Dordrecht: Reidel) p35
Wolf C J E and Rayet G A D 1867 *Comptes Rendus Acad. Sci., Paris* **65** 292
Wolff S C and Morrison N D 1974 *Astrophys. J.* **187** 69
Wolfson K L T 1980 *Mon. Not. R. Astron. Soc.* **192** 881
Woodsworth A W and Hughes V A 1976 *Mon. Not. R. Astron. Soc.* **175** 177
Wray J D and Corso G J 1972 *Astrophys. J.* **172** 577
Wright A E and Allen D A 1978 *Mon. Not. R. Astron. Soc.* **184** 893
Wright A E and Barlow M J 1975 *Mon. Not. R. Astron. Soc.* **170** 41
Wynn-Williams C G 1977 *Mon. Not. R. Astron. Soc.* **181** 61P
Yahel R Z 1977 *Astrophys. Space Sci.* **51** 135
de Young D S and Burbidge G 1979 *Nature* **281** 183
Young P and Schneider D P 1980 *Astrophys. J.* **238** 955
Zealey W J, Dopita M A and Malin D F 1980 *Mon. Not. R. Astron. Soc.* **192** 731
Ziolkowski J 1976 *Astrophys. J.* **204** 512
Zuckerman B, Gilra D P, Turner B E, Morris M and Palmer P 1976 *Astrophys. J.* **205** L15

Index

Page numbers in bold typeface indicate the beginning of a major section on the subject.